Addison Peale Russell

Sub-Coelum

A sky-built human world

Addison Peale Russell

Sub-Coelum
A sky-built human world

ISBN/EAN: 9783337037819

Printed in Europe, USA, Canada, Australia, Japan

Cover: Foto ©ninafisch / pixelio.de

More available books at **www.hansebooks.com**

Sub-Cœlum

A SKY-BUILT HUMAN WORLD

BY

A. P. RUSSELL

AUTHOR OF "A CLUB OF ONE," "LIBRARY NOTES," "CHARACTERISTICS"
"IN A CLUB CORNER," ETC.

Servant. Where dwellest thou?
Coriolanus. Under the canopy.
Coriolanus, Act IV. Sc. V.

BOSTON AND NEW YORK
HOUGHTON, MIFFLIN AND COMPANY
The Riverside Press, Cambridge
1893

CONTENTS

	PAGE
Favorably Situated	9
Abundance of Leisure	10
Marked Individuality	11
Exceptional Monsters	11
Conceit of Superior Excellence	12
Couples of Six	12
Schools generally Small	13
Instruction in Radical Morals	14
Chairs of Common Sense	15
The Average Wisdom	16
Instructed in Telling the Truth	17
A Sober Experience	18
Low, Unmeaning Language	18
Conversation Cultivated	19
Specialties of Every Sort	20
Ingenuousness a Social Excellence	20
Behavior	21
A Habit of Charitable Judgment	23
A Favorite Illustration	24
Their Superior Men and Women	25
Perpetual Surprises	26
Their Social Meetings	26

Contents

Their Floral Exhibitions	28
Floating Radiances	30
Each Day's Dinner	30
Cooking a Proud Art	33
Morals and Stomach	34
Bread-Making	35
The Papaw	35
Their Good Physicians	36
The Quack Doctor in Contrast	38
The Clergy of Sub-Cœlum	40
The Golden Rule	41
Clergymen of a Certain Character Extinct	43
The Lawyer's Office	47
A Turn was Made, but Slowly	48
Effects of the Change	50
Arbitration	52
Advice Offices	53
Laws Few in Sub-Cœlum	54
Special Reformers not in Favor	56
Effect of the Pervading Individuality	58
Their Police System	59
Fatality of Heredity	60
Estates Limited	62
Property in Friends	64
Idleness Disreputable	65
Indolence	67
Trifling for Selfish Ends	68
Ambrosia for the Soul	70
Making and Earning Money	70

Contents

Manhood and Personal Freedom	71
Native Manhood	72
Ideal Manhood	73
The Plebeian and the Aristocrat	74
The Vices	76
Common Sense and Practical Wisdom	79
Small Farms Preferred	80
Fish-Ponds	81
Bee-Culture	82
Propagation of Poultry	84
The Bird of Excellence	84
The Sub-Cœlum Oyster	85
Grapes and Wine	87
Wine-Making	88
Endless Orchards	90
Highways Ideal	92
How Cities and Villages were Laid Out	92
Drainage	94
Light and Heat	95
Public Edifices	96
Hotels	98
Bells	99
Music	101
Poets and Poetry	104
Musical Voices	106
Tight Dressing	110
A Felicity to be Well-Born	112
A Composite Population	113
Weddings in Sub-Cœlum	117

Contents

Reasons for Remaining Single 119
The First Gentleman to Speak 119
The Second 121
The Third 123
The First Lady 126
The Second 128
The Third 131
Drunkenness 134
Divorce 136
Refuges for Certain Occasional Victims 137
Retreats for Convalescents 140
Hospices for Visiting Strangers 143
Inventors and Scholars 146
Old People and Children 148
Burial-Places 149
Little Distinction in Marking Graves 151
Funerals 153
Chapels in Burial-Places 155
Motives 158
Funeral Orations and Obituary Notices . . . 159
Vocation and Avocation 161
Awed by Understanding 162
Students in Particular Lines 163
Substance of a Lecture 165
Microscope and Camera 175
Electricity 178
The People did not Snore 179
Whistling 180
Dentistry not a Profitable Profession 182

Contents

Fries Utterly Banished	183
Fondness for Squirrels	184
Respect for the Monkey	186
Instinct of Satan	190
Qualities and Faculties of the Dog	191
Horses bred for Moral Qualities	194
Beauty on Horseback	195
Love for Birds	196
Insects and Reptiles	201
Infusoria	202
Character and Mental Resources	203
The Individual the Immortal	204
Personal Independence	206
Men wiser than Sheep	208
Individuality Made Them Interesting	210
The Law of Diversity	213
The Healthful Habit of Occupation	217
The Vice of Indolence	218
Probably and Perhaps	221
A Treasure	224
The Social Conscience	224
Amusements	227
The Rule of Ezra	231
Drawing, Painting, and Sculpture	232
Not Ambitious of Great Libraries	237
Thoughts and Conduct	239
The Press	240
Results of Evolution	244
Pride of Profession	247

Contents

The High Estimate put upon Woman	248
A High Order of Wisdom	249
How Government was Supported	251
The Machinery of Politics	254
Essential Excellence of the People	258
Their Religion	262
Sects and Creeds	265
Worship	267

SUB-CŒLUM

F the people of Sub-Cœlum were not happy it was their own fault. Their situation was the most favorable under the sun. Earth and sky smiled upon them. The climate was genial and salubrious. Extremes in temperature were not frequent, and atmospheric violences so rare as to be historical. Seasons of rain and seasons of drouth, to devastate and desiccate, were not known. Forests of beauty and grandeur supplied every variety and quality of timber, for ornament and utility. Mountains of sublimity and valleys of fertility abounded. Large streams ran by large towns. Lakes bordered villages and villas. Ocean provided cities with safe and commodious havens. Gold and silver in the mountains lay in strata convenient to be operated. Where the land was poorest and least productive, the most valuable of the

<small>FAVORABLY SITUATED.</small>

<small>Silver and gold.</small>

precious metals, in nuggets, was deposited; and in the streams of such parts the most perfect of pearls, and occasionally diamonds, were discovered. Happy people! What they had not, imagination must labor to supply. Misery, to any great extent, abode not with them. So it would appear.

Pearls and diamonds.

The ease with which life was sustained left them abundance of leisure. Bent was indulged and tastes were gratified. Advantages were turned to account. Not so much to get wealth as to acquire the art of living. To make the most of themselves and to enjoy the greatest amount of rational pleasure was the common ambition. Selfishness, the one great enemy of mankind, was under perpetual ban. To gain the mastery over themselves, by studying and practicing moderation, self-control, and humanity, was the prime object of all personal and organized effort. Simplicity and modesty were at a premium, and self-respect and fellowship were exalted to a high place among the virtues. The great purpose of society was to produce genuine, individual, friendly men and women, and to surround them with all auxiliaries and facilities for growth and happiness.

Abundance of Leisure.

Simplicity and modesty at a premium.

An Object of Pity

Marked individuality was conspicuous amid all the seeming confusion it created. The great good of it was to cultivate mutual consideration. Toleration became a necessity. Inseparable from it was a strong pervading sense of justice. The right of each to be an individual man involved the right of every other to be the same. It encouraged diversity of view while it forbade dogmatical disputation. The possibility of mistake compelled generosity of judgment. Feeling was repressed and reason stimulated. The occasional man who was always right, was an object of universal pity. His deficiencies were a study and his conceits a warning. If ill-natured also, Wisdom walked by him, as by a bad animal, and Charity guarded him against irritation and abuse, — curing him finally, if curable, — eleemosynary provision being made for the worst cases.

Marked Individuality.

The man who was always right.

The snarling, venomous creature, who hated everything, and the motive-monger, who was always finding the worst reasons for everybody's actions, were the exceptional monsters — confounding the philosophers and the moralists. They skulked and they crawled, in defiance of all rules

Exceptional Monsters.

and appliances, and fattened upon their own poisonous secretions. Every populous neighborhood had one or more, to tolerate and avoid — incorrigible objects to even the most hopeful of reformers.

<small>CONCEIT OF SUPERIOR EXCELLENCE.</small> The wonder was that such characters so often had the conceit of superior excellence, which made them particularly interesting. They believed themselves better than their neighbors, while their extraordinary pretensions only made them more emphatically unregenerate. Seeing only outwardly, other people's sins exasperated them. Society, to them, was but an exaggerated reflection of their own condition. Their own moral machinery being in distressful disorder, all the world must be taken to pieces, mercilessly made over, or go to ruin.

<small>COUPLES OF SIX.</small> It was the conclusion of intelligence that eyes, to see, must be in couples of six — the pigs having two. That to see indeed — within and without and all around — eyes moral and eyes intellectual were as necessary as eyes physical. Education and conduct in Sub-Cœlum were upon that determination. It was never lost sight of.

Vanity Circumspect

It tended to make people reflective, considerate, and charitable. Self-estimates were thoughtfully made, and constantly revised. Vanity was circumspect. It was discovered that the truth, absolute and unmitigated, is hard to arrive at: that the last fact is ever necessary to correct judgment: that color depends upon light: that good is largely in the brush, and that evil is never so black as malignity paints it.

Self-estimates thoughtfully made.

Their schools were generally small, with not much system about them. No great pains were taken to force the children, especially while they were little. Memory was respected, and not over-exerted or burdened. Processes were to develop, as far as practicable, consistent with healthful growth, the best qualities and faculties of individual pupils. Differences, moral and intellectual, were recognized and regarded. It was not thought possible to make all alike, as eggs in a basket. Classes, for that reason, were limited, and specially instructed. Teachers were chosen rather for character and manners than for scholarship. Thorough gentlemen and ladies were preferred. Influence for good was looked to as a prime factor. The ready imitative-

Schools generally small.

Classes limited.

ness of the young was made the most of.
Good examples were set before them — the
best specimens of men and women procurable. Inaccurate language was exceptional in the schoolroom. The common blunders were placarded on the walls. Small children were taught by women; at eight or nine years the sexes were separated, — the girls to be instructed by ladies and the boys by gentlemen; to give opportunity, little by little, of imparting and impressing in a thousand ways a thousand things essential to genuine manhood and womanhood. To make good, intelligent, self-respecting men and women, fitted for self-government, was kept in view as the great object of education.

Good examples set before them.

In every part of the Commonwealth schools for all ages and both sexes were established, where the people were instructed in radical morals, as essential to true religion, and inseparable from it. Personal responsibility was inculcated. Marriage was gravely considered. The relations of the sexes were discussed in every way but the trifling. The nature and ethics of debt were pondered and thoughtfully illustrated. Integrity was enforced

INSTRUCTION IN RADICAL MORALS.

The Thing Necessary

impressively. Honesty to the core, in all that it implies, was persistently urged as the thing of all things necessary to true manhood and womanhood. Prudential considerations were the last to be named in connection with it.

Chairs of Common Sense were set up in the universities. Wise professors filled them. The distinction between scholarship and usefulness was continually defined. Education was directed to its uses — even to the unlearning of what could not be applied — adapting it to the character and wants of each individual — anticipating, as far as practicable, occupation and position in life. Boys were taught an apprehension of the diffusion and universality of intelligence; that no man had it all, but every man a little; that the average was always worthy of respectful consultation; that the education of the schools was but as the scaffolding and tools to the builder — bearing in mind all the time that the building that was to endure was not made with hands; that the hodman and the farm hand must teach him many things he must know; that the classics — valuable enough for culture — and the maxims of philoso-

Chairs of Common Sense.

The education of the schools.

phy must give way, again and again, and without humiliation, to the commonest experience of the meanest man, whom he would despise, till he had fairly put his mind and fact to his in the conflict of affairs; in fine, that he must surrender his self-conceit, be put upon his feet with the crowd, and totally unlearn and forget very much that he had learned, before he could begin to be truly sensible and wise.

The average wisdom. By such means the average wisdom came to be respected. It was the admitted gauge of civilization. It appeared too slow to the seer and too fast to the philosopher; but the prescience of the one and timidity of the other were not often consulted. It gave a sympathizing ear to the fervid thoughts of enthusiasts and reformers, cooling and utilizing them by diffusion. It took from the wearied eye and nerve-shaken hand of the inventor his invention, and put it to work in the fields and seas.

The common sense and the common law. It was the common sense and the common law of life. It governed the Government and every man. It put a hope into the heart, and helped it to pray as well as to work. It fostered ideas of progression, which grew into system, and methodized thought and exertion. It made tests for formulas and platforms, and widened their scope and purpose to a generous breadth and humanity. In its providence, it cared for all, the little and the great, the strong and the feeble. Its modes appeared leveling processes, but the valleys of shadow were lifted up. The sun, if it did not glitter upon a promontory, warmed the plain to produce a generous harvest. If genius seemed a little crippled in its wing, it was by teaching it a steadier flight. If the hills were less beautiful by cultivation, the vintage was compensation. In short, scholarship, less didactically and showily stated, was esteemed and urged, in that department of culture, as but a means to the end — peaceful and enlightened society, governed by humane and beneficent laws : an Ideal Republic.

It cared for all.

In the schools and universities great pains were taken to instruct in telling the truth. The viciousness of habitual extravagance in language was explained and illustrated. The close alliance between exaggeration and lying was made apparent, and all were made to feel their responsibility in speech. Volubility was discouraged. Drilling in narration was constant and uni-

INSTRUCTED IN TELLING THE TRUTH.

versal. Facts were stated and incidents related to be repeated. The practice was amusing till the consequences showed themselves to be grave. The same story, passing through several minds and repeated by as many tongues, was hardly recognized, and the result became a sober experience. It infixed itself in the memory. The dangers of careless speech, as they were comprehended, became startling. Habits of attention, therefore, and studied fidelity in repetition, were set down conspicuously among the social virtues. Truth-telling was impossible without them.

A sober experience.

The use of low, unmeaning language was considered an offense against intelligence and good-breeding, and was in every proper way discouraged and prohibited. Its rudeness and inelegance were not the only objections to it: it corrupted the carefully guarded tongue of the people. The language, in thousands of years, had grown to be so extensive that its dictionary was in many ponderous volumes. The effort for ages had been to reduce it — to eliminate all that was obsolete and impure — daring colloquialisms even being excluded. Enlightened men and women were known

Low, Unmeaning Language.

Daring colloquialisms excluded.

and rated by the purity and integrity of their speech ; standards of expression were high, and not to be despised ; rank was not risked by careless observance. Not that there was any lack in freedom of utterance. Forbidding the exceptionable encouraged the best. Intellect was not shorn of her wings. Imagination soared and gayety disported at will. Ideas, lighter than air, clothed themselves in affluent language. Humor gladdened and glowed in an easy flow of words, and wit flashed out in verbal splendor.

Standards of expression high.

Conversation, indeed, was cultivated and practiced ambitiously, but cautiously. Rude language and bad grammar were socially punished in emphatic ways, and people of good standing, making any pretensions to good-breeding and culture, were careful to be guiltless of them. Those who violated in either, whatever their scholarship, were set down as vulgar and illiterate. The general readiness and felicity were remarkable. Euphuism was rare. Affectations and excesses of free expression were instinctively avoided. Inborn taste and tact governed their intercourse. Gossip was high art. Trivialities were

Conversation cultivated.

Euphuism rare.

adorned and illustrated in a manner to create and maintain interest in them. Light philosophy turned the smallest events to account, and made each one seem important and respectable. Habits of adaptation led them into every sort of specialty. Hardly anything but had its experts and professors. Hints from nature were realized in mechanism and art. Novelties, improvements, inventions, were numberless. Every flying and creeping thing had its enthusiasts and exponents. Ephemera, infusoria, animalculæ, were classified and individualized, without limit. Microbes, bacilli, were pets of the imagination. Children, even, seemed familiar with the monsters of the microscope, and talked of them as glibly as of their playthings and the chemical elements.

Specialties of every sort.

Eagerness to know seemed not to exceed the willingness to impart. In personal affairs, secrecy was exceptional. Where acuteness was universal, discovery was nearly inevitable. Concealment being next to impossible, few thought of attempting it. Ingenuousness, perforce, became one of the social excellences. Autobiographical writing was in fashion. Publi-

INGENUOUSNESS A SOCIAL EXCELLENCE.

A Distinguishing Charm

cation of such self-revelation being in violation of the public taste, manuscripts accumulated in private cabinets, to be consulted only in social emergencies. Reporters were everywhere respected and deferred to. It was considered squeamish to withhold information from them — the reporters themselves being trusted to judge of its fitness or unfitness for publication. They made visits from house to house, and it was expected that everything of general interest would be communicated to them. Cases sometimes occurred when public indignation was aroused by efforts to misinform, divert, or baffle the indispensable news-gatherer.

Reporters respected.

The desire to behave well was as general as the desire to talk well. Politeness was a distinguishing charm. Manners were simple and easy. Stateliness was avoided. Offensive familiarity was scarcely known. Intrusion was frankly apologized for. Side-door visiting was not tolerated. Compliment was cultivated. To say pleasant things to one another was the universal custom. All were gratified by praise; they only wanted it to be sincere. Fulsome flattery was received in a way to for-

Behavior.

The universal custom.

bid a repetition of it. It was considered a cheapening and degradation of one's self to invite it, and a duty of refinement to rebuke it. Ladies set their faces against it. Sarcasm was not often indulged, and only then between close friends. When ill-nature prompted it, it was a crime against the peace of society. Obliquity of every sort was distrusted. They had a bad opinion of the lion on account of his step. Directness was preferred, even to the extent of incivility. It was a great offense to be called cunning or shrewd. Artifice was the sign of a wry mind and perverted heart. To say slyly what would occasion unhappiness was an outrage to justify punishment. Good-nature and humanity were shocked by it. To make others happy was the rule and practice; the contrary was the rarest exception. Especially it was the habit to give the greatest encouragement to worthy effort. Good deeds were heartily commended. By that means young and old were stimulated to do their best. You never met a boy or a girl who had not received encouraging words. Approbation was in every face. Hope was kept alive by it. Hearts were made human. They flowed together in good-fellowship.

Sarcasm not often indulged.

Approbation in every face.

The Difficulty of Moderation

A habit of charitable judgment had a refining effect upon the people. Experience made them cautious in condemning. They were taught to know the limits of bad and good — that nobody was quite perfect enough to merit deification, nor so utterly corrupt as to be a castaway. That a man must be looked at all around, within, by a fair light, and with a good eye, to be seen truly and judged justly. They were taught the difficulty of moderation: that if calm and deliberate enough to be just, they were almost sure to be indifferent: that ignorance, interests, prejudices, blinded their eyes, darkened their minds, and inclined them to violence. If a story came to them derogatory of a friend or neighbor, they first asked themselves, Is it true? Is it a natural thing for the man to do? Is he capable of such an act? Deliberation made them slow in determining and cautious in accepting; certain that the truth would present the matter differently. Hesitation made them charitable. It inculcated making the most of the good and the least of the bad, and to hope accordingly. They were refined by generosity of judgment, as they were made modest by introspection. Epithets of derogation and

A Habit of Charitable Judgment.

Effects of deliberation.

condemnation were rarely used. Motives were not closely questioned. Sincerity did not need to be proved. Virtue was not absolute. Intelligence, at best, was extremely limited. At sea, they said, a person's eye being six feet above the surface of the water, his horizon is only two miles and four fifths distant; yet his tongue will as freely wag of the world as if it were all spinning under his eye. We freely discuss the ignorance of those we believe to be less intelligent than ourselves, never thinking that we are the cause of like amusement to those who are more intelligent than we are. Fewer laugh with us than at us. The grades are so many that contrast is more natural than comparison. Unfortunately, too, it is only in the descent that we can see, and that but a little way. We know it is up, up, that we would go; but the rounds of the ladder are but vaguely visible. But a small part we perceive of the prodigious sweep from the lowest ignorance to possible intelligence. Upon their feet with their fellows, and conscious of the countless limitations to wisdom and virtue, the people of Sub-Cœlum grew more refined and truly polite as they became more modest and charitable.

A favorite illustration.

The prodigious sweep.

Character not in the Market

Their superior men and women were held in high estimation, and the influence they exerted was everywhere apparent. Society in many cases seemed only a reflection of them. Their high standards of conduct toned and tempered minds and hearts in remotest relations with them. The atmospheres they made and carried with them were pervasive. It was beautiful to see the respect and deference that was paid to them: silently and unconsciously paid, as the mimosa renders homage to a passing creature. Flatteries were not heaped upon them; the excellences they incarnated forbade grossness or indelicacy. The wisdom they dispensed and the good they did were not for compensation. Character was not in the market. Meretriciousness did not attempt to entice, nor artifice to purchase. Ingenuousness was a perpetual rebuke to devices, disguises, obliquities. Compliment was best paid to superiority by adopting whatever was possible of preëminence. Mere ability was not so highly esteemed as integrity — entireness. Men who were morally sound — incapable of duplicity and baseness, and women who were genuine and pure — of all excellence, were objects of unconscious

Their Superior Men and Women.

Flatteries not heaped upon them.

Ingenuousness a rebuke.

reverence. In their lives were taught virtue, honesty, honor, humanity, charity — all that constitutes true manhood and womanhood. When a superior man or woman entered any assembly, there was always more or less of sensation visible in visages and slight movement. Such personages were perpetual surprises. They were better than they appeared, wiser than they assumed, did more than they promised, and were encouraging phenomena in virtue and humanity — examples of all that is precious in character.

Perpetual surprises.

Their social meetings were all that could be desired to promote harmony and good-neighborship. They met together cordially, without awkwardness or ostentation. Manners were such as good sense and good feeling had suggested and determined. Excited and rapid conversation, as stated, was not in good taste. To talk much or eagerly was not a common ambition. Speech was upon the assumed basis of general intelligence, and was supplementary or complementary. To assume ignorance, to enlarge pedantically, were sins against good-manners; decency was offended by them. Patronizing ways were not thought of, because

Their Social Meetings.

Sins against good-manners.

not tolerated, — equality, for the nonce, being the prime condition. The happy few, with exceptional animal spirits and tact, who were able to fuse elements together, were acknowledged social forces: as moral and intellectual amalgams, they were duly appreciated; wherever they appeared, insulation was impracticable. Whatever of dexterity they employed was not easily discernible; show of management or manipulation would have been fatal. Fashion was not omnipotent, though exacting. It was hardly a device of ugliness to entrap beauty. Loveliness, in a great degree, was independent of it. Youth and beauty, in simple dresses, were conspicuous. Only the middle-aged and old dressed richly and expensively. Diamonds and gold were too common to be often used for personal adornment. Intelligence in the eye, roses in the cheek, charity on the tongue, were better than all artificialities. Figure was displayed, but not the charms of it indelicately. The consciously well-dressed were least so. Immodesty, or anything that suggested it, was not seen. Rudeness, even, blushed at the thought of it. Beautiful women were beautiful as they appeared pure. Deceitful enticement in the slightest

Acknowledged social forces.

Better than all artificialities.

Incarnate virtue ideal womanhood. made them ugly. Incarnate virtue was ideal womanhood. Men honored it above everything earthly. It was reverenced in their mothers, their sisters, their wives, their daughters; and their treatment of all women was touched by the distinction. In their social parties both sexes of all ages commingled — a few children being considered necessary to a complete company, as undergrowth is indispensable to a healthy forest. *Respect and amenity.* Respect and amenity characterized behavior and word. The young were deferential to the old, and the old considerate of the young. Venerable ladies received the attentions of young men, and venerable gentlemen extended every politeness to young women. Age and youth were side by side in the dance and at the banquet. Courtliness and the small sweet courtesies were taught and practiced. Manhood was improved and womanhood exalted. Human nature appeared best in the brightest light. Pessimism, even, if it existed, thought it worth while to continue the race under hopeful conditions.

Their Floral Exhibitions. Not the least attractive feature of their civilization was their floral exhibitions. The universal taste and a generous rivalry made them frequent. Everybody attended them, and the enthusiasm shown was beautiful to see. Men and women had become famous by cultivating and propagating particular species. Gardens of roses and gardens of pinks were everywhere. Varieties seemed infinite. The bloom of the dandelion and daisy was grown to be thrice as great as in the wild state. The hollyhocks were prodigious. The geraniums blazed in a marvelousness of color. Chrysanthemums of bewildering variety and beauty were the pride of the multitude. Pansies appeared living creatures. In these shows the best achievements in floriculture were brought together. The taste displayed, and the abounding beauty, made them delightful and memorable occasions. But more attractive than the flowers were the throngs of humanity that moved amongst them. Beauty was made more beautiful and nobility more noble by being brought together so auspiciously. All that was good in man and woman seemed to shine out in happy faces. Roses in cheeks bloomed with a warmth the roses in the gardens did not have. Expression was animated by the enlivening scene. Beauty was surprised into attitudes that poet or

Gardens of roses and gardens of pinks.

Beauty and nobility brought together.

Two floating radiances.

painter had never witnessed. The two floating radiances that appear and disappear amidst the roses! Noiseless as spring sunshine and as inspiring. Blonde and brunette, distinct, together, and blending. Raven hair and golden, rippling at random and flowing together. Blue eyes and black, alternating; confusing your fancies, like the changing hues of a sunset. Complexions nut-brown and alabaster, warm and roseate with innocency and ripeness. And

The good woman of threescore.

the good woman of threescore who exchanges civilities with them! Her complexion is as clear and her face almost as sunny as theirs. That glistening silver lock must but a moment since have turned gray while she unconsciously twisted it. Her voice and smile and eyes do not answer to so much of life and vicissitude. The three sympathize and mingle, without adjustment or dissonance. Happy children and grave men add to the diversity of the occasion. What could be more elevating, picturesque, or wholesome, than human intercourse under favorable auspices?

EACH DAY'S DINNER.

Each day's dinner was much of an event in every family. It came early in the afternoon, as the hours of labor and business

The Family Dinner

were not many. It was the rule to forget the cares of the day, and to put away anxiety, as far as possible, in preparation for it. Plenty of time was taken, to fully enjoy it. Not that the population were especially devoted to eating; they looked more to the civilities and socialities than to the indulgence of the appetite. Cleanliness was particularly observed, in person and in table-habits. Promptness was expected of every one, and a careful consideration for the comfort and pleasure of others was maintained. Each one took his place, without eagerness or disorder. The service was deliberate, and in courses — chemically right foods being served together. Tables were padded to limit the noise of dishes. Personal peculiarities of taste were ignored or not referred to. Noise in eating was scrupulously avoided. Pigs for that, they said, not men. Children were so instructed, but not at the table. Rapidity was not indulged, for the same reason. A famished manner was offensive. Excess in quantity also. Repletion was as objectionable as voracity. The dishes served, their costliness and preparation, were not elaborately discussed. Dining was else and more than feeding. It in-

The civilities and socialities.

Peculiarities of taste ignored.

cluded all that was civilized and generous. Best impulses were quickened and liveliest ideas evolved. Irony was not indulged at the expense of good-nature. Good feeling was requisite to a good dinner — a better sauce, if possible, than hunger. Words were not taken from others' mouths; interruption was rudeness. Subjects introduced, as far as practicable, were elevating, but not above the range of the average. Free utterance was encouraged, but not, as before observed, too great precipitancy or volubility. Discoursing, or talking in a lecturing way, was a violation of good table-manners. That every one might have due opportunity of participating, anything like monopoly, if indulged, was jealously restricted. Children were encouraged to a full share in conversation. Occasion, indeed, was often made to give them prominence — self-instruction being an ulterior purpose. Birthdays of distinguished men and women were selected for their special benefit. A little better dinner than usual was provided, an extra dish or an additional course being sufficient. A suitable guest was selected to partake, and to put all upon their good behavior. The children were expected to lead on these anniversaries.

Good feeling requisite.

Children encouraged.

Ample time was given them for preparation. Dictionaries, cyclopædias, and biographies were consulted for facts and incidents. Each one was depended upon to contribute an anecdote or interesting fact. Contemporaneous history was recalled. Lessons in philosophy and conduct were suggested. The good in the several characters considered was brought out exemplarily, and the bad referred to in admonition. The great and excellent in life and literature were thus studied and kept in memory. All were made to think, and to grow in enlightenment. The children especially were helped and stimulated in self-education.

Lessons suggested.

Cooking was a proud art in Sub-Cœlum, and was carried to great perfection. Still they experimented, and their best results were from time to time announced in gastronomic journals. The invention of a new dish gave distinction, next to the discovery of a new planet. Chemistry was so persistently and ingeniously applied that kitchens became laboratories. Bad cooking was a sin, and brought shame upon the sinner. This extraordinary interest in the art was due in great part to the prevailing

COOKING A PROUD ART.

Morals and stomach. opinion that morals largely were emanations of the stomach, and that men were good and healthy as they were well fed. Curious and wonderful instances were collected in proof. Crimes were traced to bad breakfasts, as benefactions were to good dinners. The philosophic cook accounted for conduct as he did for complexions. Roses in cheeks told their history. *Sallowness a reproach.* Sallowness was a reproach, and was very rare. The shades of melancholy appeared in few faces. It was the general belief that most diseases were caused by bad or ill-cooked food, and that few of them that were remediable would not yield to right diet. The doctor often, before writing his prescription, questioned the economy of the kitchen. The priest, before consolation or absolution, did the same. Courts, in the trial of criminals, directed similar inquiry, and extenuation or commutation was often a result of it. Law-makers were indebted to cooks for suggestions. Moralists were liberal as they were gastronomically wise. Pork was held accountable for *Roast pig.* much that was bad in the world, roast pig excepted. The young of swine, something heavier than a full-grown capon, were objects upon which genius expended itself. The sweet juices thereof reached the sources of sense, and remained in the mind as on the palate, inclining it to generous reflection. Fish, too, the particular *Fish.* food of the brain, employed and exhausted the possibilities of kitchen science. Never a drop of water entered into one of the finny tribe after the knife had done its office. The natural juices were all preserved — every particle. Banquets exclusively of fish, with ichthyological pictures all round, were not uncommon events. Symposia they were of wit and eloquence. Bread-making was carried to great perfec- *Bread-making.* tion. Loaves were congeries of sweet crystals. The light shone through them. They were marvelous. Common articles were made wondrously palatable by the manner in which they were cooked and served. Fruits especially were temptingly presented. The papaw, the North Ameri- *The papaw.* can custard apple, was a favorite of the people. It was sedulously cultivated, and was considered excellent above all other fruit. Ripening upon the tree, and falling upon the leaves, it caught a taste of earth and heaven that was ambrosial. It was the supreme delicacy, and was daintily eaten. Nothing so palatable, they said;

certainly nothing uncooked. An appropriately artistic dish received it. The knife to lay off its skin was set in diamonds. The spoon to eat it with was of purest gold, of delicate and exquisite workmanship. A bit at a time was enough, every atom of which rose to the sensorium. A half an hour was considered too short a period to linger over this achievement of nature — her one inimitable, unsurpassable custard. The beautiful orchards of this Fruit of Paradise were the triumph and pride of pomology.

A bit at a time enough.

The people, being highly intelligent, required the best of physicians. A little smattering and a great deal of pretension would not do. Ignorance, that presumed to exercise important functions, was held to be criminal. When exposed, it became an object of public reproach. It might trifle with anything but human life and not be declared odious; but when poor human bodies were subjected to merest empiricism, the public sense and the public conscience revolted. It was understood that the more knowledge the physician had, the better fitted he was for his profession. No man, in their judgment, could know too

Their Good Physicians.

much to be a good doctor. There was, they knew, no end to the knowledge applicable to the treatment of disease. The physician was not expected to perform miracles, as the world had grown too wise to expect the miraculous. The human machine was admitted to be frail, and destined to go to pieces. The house of clay was only to be kept in such repair as to be presentable and comfortably habitable till abandoned. It was not made to resist earthquakes nor time. Only the everyday storms and ills were to be averted or cured. The one great shock or poison, which shatters or rots the structure, the wisest could not forefend nor baffle. Therapeutics could not be so exact as anatomy. With the aid of anæsthetics, the medical carpenter might cut and saw his poor fellow with certainty. But the many influences of climate, appetite, and passion upon human bodies, as varied by predispositions, habits, and ambitions as they were numerous, were admitted to be past finding out mathematically, and too often were only to be guessed at, as the turns of the market, or the whims of insanity. Sincere, and devoted to his calling — carrying conscience into it as well as intelli-

The house of clay.

Influences of climate, appetite, and passion.

gence — the physician was not expected to accommodate himself to pretenses nor whims. A professional call meant sober business, and his sense of duty commanded candor. If indolence, or indulgence, or vice were the cause of ailment, he frankly announced and characterized it. The cherished habits, appetites, or desires must be abandoned before he could begin a cure of their results. Describing their effects upon the body, he did not suggest their blighting consequences upon the character. That, he considerately left to the curer of souls — the clergyman or the priest.

A professional call meant sober business.

Time was, even in Sub-Cœlum, when the quack doctor — the empiric, the charlatan, the pretender — was in fashion. To appearances he was most considerate and respectful, while with the real he made merchandise. A large proportion, and the most substantial, of his patients, were only growing old, but they submitted to be drugged and drugged, rather than to be told the wholesome truth. The slight weaknesses and aches, as natural as gray hairs and dim eyesight, pride of life and the pretender's arts dignified into illnesses. Thin locks and spectacles were natural

The Quack Doctor in Contrast.

Dignified into illnesses.

Wise Prescriptions

enough, and well enough, and becoming; but flattening muscles and cooling circulation were results of over-work or imprudence, and might be restored to roundness and comfortable temperature. The doctor's wise prescription was higher living and heavier woolens, with powders and drops now and then as alteratives and tonics, and just soon enough, to a visit, he conducted the case to a favorable issue. The air of another clime was recommended if his patient's patience seemed failing, or if, as the real case might be, the ill-judgment of a stubborn husband was to be corrected. Many of his patients who were given to gayety and irregular hours, who were too frail to bear children, his mere hint of the fact was of profit to the monster in a palace, whose specialty was such cases. Expressionless eyes and dullness contrasted with beauty and thin dresses, and stimulants in every form were suggested to supply the needed lustre and sprightliness, and complete the harmony. Small potions at first were sufficient; and if gradual increase of quantity resulted unfortunately, the misfortune was disease, to be treated by a still further increase of the cause as a remedy. If the public voice

Heavier woolens.

Small potions at first.

Crime in many households. was silenced by the presence of crime in so many households; if brothels sprang up palatially in desirable streets; if hospitals multiplied to exhaust the public purse; the fashionable doctor, who was the genius and patron of them all, was secure in his fame and opulence. Long, long since, the people of Sub-Cœlum had grown too wise and excellent to tolerate such an embodiment of insincerity and artfulness. They preferred conscience and candor in their physicians, and profited by them in body and in spirit.

The Clergy of Sub-Cœlum. Time was, also, when the clergy of Sub-Cœlum were not all that they might have been. Too many of them had grown worldly and time-serving. The occupant of the carved pulpit, it was said, whose wants were only imaginary, knelt upon cushions of velvet, and thanked gracious Heaven for having made the circumstances of all mankind so extremely happy. Material demands upon him having been paid by checks on his banker, he was profoundly *Ignorant of the multitude.* ignorant of the shifts of the multitude. Here and there, in pews nearest the pulpit, reposed, in fresh raiment and elegance, representatives of every institution of finance and commerce; and their joint possessions impressed him with the fullness of beneficence. To illustrate his theme, he was not limited to average experience, but was expected to range beyond and above it. He was understood to know the world in an enlarged way; and if his figures or examples suggested the successes or power of certain of his hearers, their complacency was stimulated if their hearts were not softened. He was not to shock by an exposure of subtlety which circumvented, or combination which oppressed, but to soothe by a glittering exhibition of ends and attainments. The possession of money, in whatever prodigious quantity, was not to be questioned, but only the love of it. A little ingenuity would comfort the possessor by suggesting his expenditures, and make him as conspicuous in the sanctuary as his equipage made him in the avenue. His thoughts were especially pitched to the ears of those he besought, to whom he owed all, and from whom he expected even more. Their courtly presence he had enjoyed till their moral atmosphere had become his own, and his passions flowed much in the same current with theirs. Once a year the Golden Rule was dwelt

Not limited to average experience.

Love of money.

The Golden Rule.

upon, to harmonize and conciliate commercial niceties. Refinement and speciousness might display themselves upon so sober a generality. If not a vague abstraction, it might be only relative in its application to life, as honesty in the common sense was not integrity. Honor was the practicable and necessary rule. The speculator might have it, and trade by it, though his ingenuous friend be ruined by his scheme. Thief he might morally be, and a beggar his dupe, but the contract must be fulfilled, and justice indorse it. As the ethics to govern in the settlement between man and his Maker, with character only in judgment, the Golden Rule was unquestionable, but not in the court of the moneychangers, where honor alone gilded the edges of promises. With the motives of the heart God must deal. Man must have his due. He was a pretty preacher for young people. His manner moved them like the bursting spring. His similes were of buds, and blossoms, and fresh verdure. His soft words and gentle gestures winnowed fragrance. His accuracy and variety of taste made him a connoisseur in everything pertaining to colors and fabrics. At home everywhere where there was ele-

Honor the rule.

A pretty preacher for young people.

Artist and Arbiter 43

gance, contrasts and harmonies had trained him to refinement of observation, and he was at once the artist and arbiter in perplexity. As a relaxation from labor, and to gather resources for the entertainment and instruction of his people, he had traveled the world, and seen edifices, and pictures, and costumes, and his perception of effects was acute and unerring. His indorsement of the style of a house, or the beauty of a landscape, or the trimming of a gown, was assurance of grandeur, or harmony, or tastefulness. His ethics in the pulpit and æsthetics in the drawing-room were alike acceptable and infallible. Degrees of future happiness, he believed, would be determined by development. Bliss was only relative. Enjoyment, as a rule, was measured by capacity, and incongruity would mar heaven. His theory of fitness and likes solved eternal justice and harmony. *His perception of effects.*

Ages had passed since clergymen of that character had been seen in the pulpit. Extinct, they were read about as strange curiosities, and the people were amazed at the moral standards of their ancestors. That the materialities had governed them to *Clergymen of that character extinct.*

such an unlimited extent was marvelous.

A religion of Mammon. They could not comprehend that a religion of Mammon had ever existed in their fair land. The conditions of society had completely changed, and their views of worship had changed with them. Wealth being diffused, other standards of excellence and conduct had been established. Men who counted their wealth by millions were not known amongst them. Churches were not built and maintained by the prosperous only. One class was as conspicuous in their management as another, and each contributed its full share to supporting them. In their clergymen, purity above all things was a requisite. The preacher must first be a thoroughly honest man, beyond even the suspicion of duplicity or dissimulation — faithful to all of his vows. His relations with society must be absolutely

A vicegerent of God. immaculate. A vicegerent of God, he was not for a moment to forget the responsibilities of his office. He must be utterly incapable of soothing the conscience to sleep by noxious sympathy or advice; of extenuating any corrupting desire; of concealing any wholesome truth; of excusing hypocrisy in any of its multiplied forms; of countenancing philosophies to flatter worldliness; of confusing worship with ceremony; of courting power, or forgetting to enforce its accountability; of helping to degrade integrity to the standard of commercial honor; of exalting money, or disregarding improper means of obtaining it; of encouraging wine and denouncing drunkenness; of extolling prodigality and deploring bankruptcy; of magnifying costly raiment and bewailing demoralization; of cautioning youth, with only manhood, against marriage, and warning him of the strange woman; in a word, he was continually and persistently to set his face, and exert the utmost of his personal and official influence, against everything corrupting or degrading to man or woman. Not that he was to be harsh or hard to any human creature. That he might be generous in his judgments, and sympathizing towards weaknesses, he was required to be thoroughly educated as a physician before entering upon his sacerdotal office. Moral ailments of every kind would present themselves to him, and he must be as capable as possible of treating them. To account for mental diseases he must know bodily derangements. Body and mind were to be considered one and inseparable. The in-

Integrity not to be degraded.

Not that he was to be harsh or hard.

Body and mind.

terdependence was not to be forgotten. Inheritance lessened accountability: the Maker only knew to what extent. The preacher did not set up standards of conduct unattainable by himself or any of his hearers. He did not turn the key of heaven against himself and all mankind. He did not preach an empty heaven. He believed that no man was so bad but that there might be some good in him, and that no man was so good but that he might be better. The differences between the good and the bad, which, at first blush, appeared to him so great, as he knew more of man and men — more of the weaknesses and distresses and ignorances of his fellows — seemed less and less to him; and he reflected how, in the eye of the Maker, who knows everything of every one of His creatures — every besetment and every infirmity — how impossible, with all his efforts, to accomplish very much — how next to impossible to use at all his imperfectly developed wings — the good and the bad must appear pitifully alike, if not the same. His learning, observation, introspection, and reflection made him charitable. His religion was love. Hence the typical preacher of Sub-Cœlum.

Standards of conduct.

The good and the bad.

Pitifully alike in the eye of the Maker.

The Devil's Confessional 47

Time was, too, when the lawyer's office was named the devil's confessional. Whatever it was in fact, the low tone of morals was responsible for it. The long continued universal greediness in money-getting had debauched the public conscience, until integrity had come to be of inconsiderable importance in the market. Lying was excused as a necessity. Adulteration was not a crime. Duplicity had no bad name. Shrewdness was a virtue. Villainy was not such when it succeeded. Being found out was the dread and devil of the population. Concealment was studied as a refinement in business, and craft was exalted to a chief place with wisdom. Straightforward ways were at a discount. Honest poor men had no fair chance. In the combinations they were left out. The carcass was appropriated, leaving only the bones. Vast schemes were but conspiracies, — powerful enough to suborn, if they did not crush. Pettifoggers of every grade were their instruments. Rodents sometimes in their processes — angels of light at others. Bold enough to assault or corrupt at the top, base enough to undermine at the bottom. Weaknesses and tastes and ambitions were estimated as commodities, and

The Lawyer's Office.

Craft exalted.

Bold enough to assault or corrupt.

prices put upon them. Great interests at stake, equivalents corresponded, in money or advancement. Great attorneys were found to employ mean and bold arts in high places. If they failed at one time they waited for a better. A new trial was had, and the indispensable testimony was supplied. The courts being reputed corrupt, corrupters had full employment. Honest lawyers were forced to low fees and a low station. The determination to do right consigned them to poverty. Judges were welcomed in disreputable and doubtful places. They were shown the way to great bargains. They enjoyed the chances of great gain without pecuniary risk. They were in the councils of conspirators. They sailed the dizzying maelstrom of fashion, and moved omnipotently in the mysteries of markets and corporations.

Mean and bold arts.

These things could not continue and society exist. A turn was made, but slowly. The flesh was ready to fall off the bones. Constitutional remedies were applied. Little by little the moral sense was elevated. Eyes were gradually opened to the danger. New standards were set up. Reformation began at the bottom — the

A Turn was Made, but Slowly.

foundation of the social structure — and worked upward as the corrupted tissues would bear. Better blood came with better morals. Conduct found higher aims. Money was no longer the standard of excellence, nor wealth the omnipotent distinction. Intellect and purity no longer submitted to be graded, averaged, and appropriated. The virtues were at a premium. Honest poverty ranked with comfortable competency. Mammon was dethroned, and not a god. All this did get done, but it took ages to do it. The devil fought desperately for his advantages, but was routed, in person, — his creatures remaining in the trenches to make a show of resistance to virtue ever and ever. The corrupt judges had to go, with the hypocritical priests. They died hard, but they died dead. Intrenched as they were by their evil practices, society at large felt a throe of relief at their final discomfiture, notwithstanding the many who suffered by it in their interests and occupations. The vice of courts had poisoned the streams of trade to such an extent that traders traded with vicious impunity, confident, if found out, that the same vicious means would save them from disgrace, if not from punish-

Better blood with better morals.

The vice of courts.

The price of protection. —ment. The price of protection was graded to every depth of purse. The petty thief felt as secure as the great swindler. For a consideration the jury could be made to please him. His attorney would attend to that as to everything indispensable. All was made easy to him by the laws, framed by lawyers — they seemed indeed to have been made for his special protection.

EFFECTS OF THE CHANGE. When the change began in the public morals, attorneys became more or less objects of suspicion. In time, it became possible to disbar them for dishonest practices. Before being admitted to practice, their characters were scrutinizingly examined — integrity being held of greater importance than legal learning. Conversations with judges about cases, outside of courts, ceased to be common. Every word addressed to a judge by suitor or attorney must be in open session, — in a voice to be heard by every bystander. A violation of this rule was an offense against the dignity of justice, and was punishable. Desire to be a juror was proof of incompetency. No *Intelligence not discriminated against.* panel was completed without inquiry as to that; nor was intelligence discriminated against. Lawyers were no longer preferred

The Judiciary

for lawmakers; when chosen to legislative assemblies, they were of conspicuously high character. The judiciary was jealously guarded. Judges were elected for long periods. Solicitation disqualified them, even for being candidates. Names of persons suggested for judicial offices were published for a time in separate lists, and each one was carefully canvassed by the public. If any serious moral defect was discovered, the possessor of it was declared ineligible, and unfit to be voted for. Legal ability was duly considered, but not to the same extent as personal incorruptibility. The trouble with society had been that in a general way too great a disproportion had existed between those intelligent enough for places of trust and those possessing the essential moral qualities. Honesty and purity, consequently, came to be necessary and absolute prerequisites to the judicial office. Judges must be non-residents; in other words, they were required to sit in districts other than those in which they were elected. The population would not tolerate resident judges. They were thought to be too familiar with the people and their affairs, and apt to be warped in their judgments. Strangers were preferred,

Judges elected for long periods.

Honesty and purity prerequisites.

Strangers preferred. who knew nothing of society or its influences. From their places on the bench they did not look down upon suitors whom they knew intimately, and whose interests they could scarcely judge of impartially. Only blind Justice could hold the scales evenly; Mercy was an independent power, and must be consulted apart. At that court, eyes and ears were ever open to the tears and appeals of humanity.

ARBITRATION. By arbitration was a favorite mode of adjusting most of their difficulties. Adjustment being necessary, the most direct way to it was chosen. Advantages were gained by it. Delay was avoided and expense saved. Anxiety was reduced to the minimum. Time was not allowed to deepen distrust into hatred. The peace of society depended upon the promptness and thoroughness with which differences were settled. Business difficulties were adjusted by business men. Parties tried their own cases. If the laws of evidence were sometimes relaxed or overlooked, it was to give *Scope and freedom allowed.* those most interested greater opportunities to show themselves. In the scope and freedom allowed, arbitrators saw behind the faces of transactions and suitors. Ignorance was enlightened and malice disarmed by the clash of interests and passions. Motives dropped their disguises, and truth was conspicuous. Unconsciously, often, the sources of trouble were exposed in a way to make adjustment easy and unquestionable.

Advice offices, here and there, throughout society, were established. There was use for them, and they were freely used by the people. They were sanctioned and protected as were other places of business. Men of good sense and of good health were the counselors — astute of observation, and sagacious in the ways of the world. Stupid people, and people of questionable character, were not tolerated in the office. Advisers were generally well-to-do persons, and charitably disposed. To relieve and help in common extremities was their sworn duty. Fees were entirely voluntary. A misuse of the generous license given them was visited with prompt condemnation by the public. The office of a public adviser was held in sacredness next to the cloister of a priest. Poor men and women who did not see their way clear to invade the latter were accommo-

ADVICE OFFICES.

Fees voluntary

dated in the former. They were shown the way out of ordinary trouble, and encouraged to better progress. They were warned of the consequences of evil habits.

Industry inculcated, and frugality.
Industry was inculcated, and frugality. Self-dependence was impressed upon them. The pleasure of vice and pain of virtue were set down to ignorance. If their troubles were of a business nature, they were advised to arbitrate them. If a diseased condition of body or mind showed itself, they were recommended to the physician. All who came were encouraged to attend the Public Schools of Morals, and be taught the foundations of good conduct. The clergyman was recommended, or the priest, in peculiar distresses of the soul.

LAWS FEW IN SUB-CŒLUM.
Laws were few in Sub-Cœlum. Such as existed were necessary, and were strictly enforced. Their book of statutes was comparatively a diminutive volume, and there was not a dead one in it. The people did not need to be much governed — in the main, they governed themselves. Experience had taught them that laws easily executed were hardly necessary, and that those which could not be enforced were worse than useless — they were vicious.

Self-Enacted and Inevitable 55

Such legislation as was indispensable was, in a sense, self-enacted and inevitable;— in other words, was so generally required as scarcely to be disapproved. Before enacting a law, lawmakers inquired, Can it be enforced? Is society ready for it? They did not think that men could be made temperate and virtuous, or women chaste, by statute. Moral power was considered a better force than the most efficient constabulary. The disposition or desire to do wrong was before all prohibitory enactment. If that existed generally, a small minority were powerless to punish its consequences. Penal legislation, with that view, was not difficult — it was but the spontaneous expression of the multitude. The difficulty had been that Government had attempted the impossible — making itself ridiculous by empiricism. The people became weary of chimerical experiments — of all efforts to adapt them to imaginary, super-celestial conditions. They were not to be made over violently. The tiger's tooth was not to be eliminated in a generation; the slow processes of breeding and gentleness could only be counted upon in anything so radical. The habit of resisting evil was found better than all threatened reformation.

Lawmakers' inquiries.

Weary of chimerical experiments.

The claw yet in the soft paw.   For thousands of ages Sub-Cœlum had been a part of the inhabited universe, and had grown to be what it was by the slowest progress. The claw was yet in the soft paw, and was not to be torn away forcibly. Savagery was not out of sight in their civilization. Their laws were mostly to assist voluntary efforts in right directions. To aid, and not to compel, was their prime object. The edifice of their polity was of composite construction, wherein by degrees were appropriated and incorporated such elements as had been proven necessary to the safety and permanency of the structure. Ages had gone by since the people had tolerated empiricism or charlatanry in government. They would not be tinkered with or unduly agitated. Repose they esteemed a prerequisite to healthy growth. They discouraged the spasmodic, and were not ambitious of an interesting history. Health and genuineness and purity, in their judgment, were not turbulent or theatrical attributes. The universe made no noise.

SPECIAL REFORMERS NOT IN FAVOR.   Special reformers were not in favor; indeed, there were few of them. Society was so individualized that there did not

Diseased in Some Way

seem to be use for them. Such as there were, for one reason or another, were distrusted. It was observed that too often the evil they meant to correct was more in themselves than in others. In some way, from some cause, they were diseased, and the reforming spirit was a result of their condition. Healthy persons did not exhibit it. Only where the body was peculiarly afflicted, deficient, or deformed, or the mind had lost its nice balance, was this uneasy tendency inclined to show itself conspicuously. Reasoning from the special to the general, they concluded all to be in their own condition. The morbidity or painful self-consciousness that distressed themselves they believed to be pervading. The common effect was to excite pity in the sound of mind and body, and to suggest the propriety of guardianship over them. Narrower and narrower they became as they traversed their remorselessly strait and ever-narrowing path. Their own standards must be the standards of the universe or the universe was all wrong. As Philosophy said, they neglected their own fields, and went to weed the fields of others. Also, that Virtue did not take pupils; she contented herself with sowing

The reforming spirit a result of their condition.

Suggesting guardianship.

Goodness and experience. the seeds of goodness, certain that experience would make them grow. It was observed that where a disagreeable or unsightly deformity existed in the body, an answering one was apt to show itself in the character or spirit. Perpetual consciousness of it occasioned diseased sensibility, and excited a feeling of separation if not antagonism. Never forgetting it for a moment, they naturally misapprehended everybody about them.

Effect of the Pervading Individuality. The pervading individuality, as said before, made men distrustful of radical reformers, as introspection made them wisely observant and generous. It did not incline them to make others over, or to have it done. Self-reformation was a natural effect of it — the kind they thought to be, of all others, the most genuine and permanent. It led to special investigation rather than to general reformation. It disinclined them to be organized into parties — to be merged into multitudes. *They moved not in mobs.* When they moved, they moved not exactly in mobs. Leadership was temporary, and only when necessary. Then, they did not follow as sheep. They did not study to conceal their personal traits; only to train

A King Each One

them to usefulness and agreeableness. They were not made to bore, to offend, or to bully, but to make the possessor of them more interesting and serviceable. Their language was not for concealment but for expression. They meant that whatever was peculiar in their nature should not be hidden, but laid open, and turned to account. A certain sacredness was made to attach to it accordingly, as contradistinguishing each individual. He was made to do something, and to do it better than could any other. Men were not so much mysteries to each other as wonders. Each one stood forth a man, different from all other men. Recognition begot respect. Men were not to be compounded or melted into masses. A king each one, he was respected in his sovereignty — over himself.

Language not for concealment.

Their police system was inseparable from their society as organized. It pervaded and permeated every part of it. Every individual and family and organization was exposed to it. In truth, there was little of what might be called private life in the entire Commonwealth. The habits of the people discouraged if they did not forbid privacy. Their remarkable individualism,

THEIR POLICE SYSTEM.

ingenuousness, and perception — almost prescience — revealed all and saw all.

Mind and conduct reading. Mind and conduct reading had reached such perfection that wrong-doing was nearly impossible. Blinds at doors and windows were not so much to elude observation as to exclude and regulate the light. Language, as before stated, was for expression, not for concealment or dissimulation. Masks of any sort only invited inspection. Faculties were sharpened by them to microscopic accuracy. Utmost apparent candor was often more deceptive, as motive-

Motive-mongers in ill-repute. mongers, in the ordinary way, were in ill-repute. People were expected to be truthful. Falsehood was in violation of all their training. Truth was at the basis of their practical religion. Their morals reacted on their bodies. They lived to great age in consequence. By the mere power of enlightened will it seemed they lived or died at pleasure. Disease was prevented by foresight, inoculation, or vaccination. Parents, when they punished their children, were particular at the same time to punish

The fatality of heredity. themselves; — recognizing the fatality of heredity — the responsibility of paternity — that the child did not beget itself. Only murder was punishable by death. Breach

Means of Protection

of trust, ranking next in criminality, was punished with great severity. For third offenses, of any serious character, imprisonment for life was the penalty, that society might be protected, and children be not begotten by incorrigible criminals. The sins of lust were especially punished, as being radically demoralizing. The face of meretriciousness was not only a warning to the police; it was sadly shocking to decency and the moral sense; and admonished special guardians of the social superstructure to look well to the foundations. Purity, of all things, was most jealously guarded. The incorrigibly impure were locked up forever. Men and women, as to that, were treated alike by the police and by the courts. If society was to continue to exist, and grow in essential excellence, chastity must be increasingly recognized as the crowning virtue. Education, experience, hope, all inculcated it. The regulations of society were such that many opportunities for crime did not exist. The bad were found out, and thwarted in their evil purposes. Persons removing from one part of the country to another were required publicly to announce and register the same, with the causes thereof; and

Penalty for third offenses.

Purity jealously guarded.

The bad were found out.

those removing into a new community were also required to state and record in the same public manner the cause or causes which prompted their change of abode; at the same time and in the same way, to give a history of themselves — their occupations, purposes, circumstances — everything, in fact, in which the people were understood to be interested. As little as possible was left to curiosity or doubt. Men and women were known and read by all. Places in society, in a measure, were self-assigned. It was not possible for any one to be far deceived. Self-regulation was a large part of the business of society. Police officers had little to do: about all was done for them. Personality and conduct stood out so conspicuously and significantly as to make official interference only occasionally necessary.

Change of abode.

Estates were generally small in Sub-Cœlum. Great wealth was not considered desirable, and was discouraged by the population in every way that was proper and neighborly. It gave distinction not in harmony with their established system of government. Only the utmost equality was thought to be consistent with pure

ESTATES LIMITED.

democracy. This central principle was never lost sight of in all their legislation and social regulations. The spirit of agrarianism did not show itself amongst them: their singular integrity repressed it. Public opinion rather than the law fixed the burden upon property worthily, and rich people realized and accepted it. It was but the price and responsibility of prosperity. Beyond a certain limit they were taught to hold their property in trust for the benefit of the public, and of individuals less prosperous than themselves. They distinguished themselves by their generosity. Their benefactions made them popular as well as famous. Hospitals were built by them. They busied themselves quietly in searching out misfortune and relieving it. They made humanity and self-sacrifice fashionable, but not ostentatiously so. The good they did was by few words, and not by formal announcement. It showed itself rather in results. Wealth did not array itself offensively to simple livers. Socially, it kept within the average. Their banquets were not insulting in their splendor. They did not endanger pedestrians with their hurrying equipages. Their advantages were not aggressive. It appeared

The burden upon property.

Humanity and self-sacrifice fashionable.

a noble thing to enjoy opulence in a right way. Envy was not disturbed nor hatred awakened by its privileges and pleasures. *The virtues were common possessions, and disported themselves, in a sense, in palaces as in cottages. Money, in itself, did not give honorable celebrity. Distinctions of God gave greatest prominence and eminence. A man might be great, without skill to advance himself, or cash to help his fellow. The riches of heart and intellect enjoyed just estimation.

The virtues common possessions.

PROPERTY IN FRIENDS. The property of all, however — the property that ranked highest — was the inestimable property in friends. The man enjoying the greatest number of good ties was the man supremely rich. His riches were above and beyond robbery. His friends were wealth imperishable, while he deserved them. The common ambition to possess this incomparable wealth had a stimulating and exalting influence. It was property within the reach of all, and a disparagement not to possess it. The significance of friendlessness was duly estimated. It meant unworthiness, and a lack of the genuine virtues of humanity. Courage was wanting, and fidelity. To have no friends

Significance of Friendlessness

was not to deserve them, and the situation was pitiable. Utter selfishness or degradation only accounted for it. If the creature had done any generous thing, the beneficiary would have adhered to him. If he had divided his loaf, the satisfied appetite would have kept him in remembrance. If he had shown a poor man out of his extremity, the happy relieved fellow would have given him his heart. If he had been kind to children, he would have enjoyed an ever-increasing harvest of good wishes. If the old and the feeble had been helped by him, his ears would have been filled with their benedictions. If poor woman, with all her troubles, and his own too, had been met more than half way by his sympathy and tenderness, a friendship immortal would have attached to him inseparably. To have no friends was destitution indeed; but to deserve a multitude of them was to enjoy riches incomputable and imperishable. Such standards of wealth and worth were the result of experience and every test, and were fixed and irreversible.

Selfishness or degradation accounted for it.

Labor was so honored that sheer idleness was disreputable. Every one was expected to have something creditable to do,

Idleness disreputable.

and to do it. Children were brought up to pursue some avocation, or cultivate some taste. Occupation was considered an indispensable duty in the social man. An absolutely idle citizen was but one remove from a knave. To work with his own hands was not only the duty but the pride of every capable person; and prejudices which despised labor were positively unknown. Business descended from father to son, and perfection was attained in every branch of it. It was found that a man was a better bootmaker from having descended through a long line of bootmakers. The feet of one in the care of such an artist were insensibly comfortable. His brain was not racked nor his nerves tortured by a distressing localization of his sensibility. Happier, too, was the artist or artisan from perfectly understanding his occupation; and he was esteemed accordingly. A better feeling was established in life by experience of its utilities. Jealousies and envies and hatreds were restrained by it. Fraternity was made easy, and fellowship possible. Manhood was helped upward by it, and ennobled. Mere living was not considered a worthy object of life. True life was above the means which sustained

Occupation an indispensable duty.

Mere living not a worthy object of life.

it. Equanimity had an eye to results beyond the moment. Only the beasts that perish were contented to be merely fed. The nervous tread of a true man meant more than movement; it betrayed absorbment, and looked to an end worth attaining. Idleness had every gait, and none long. Whim changed it. Nothing to do was held to be the worst want of nature, and the most exhausting. It tested severely mind and morals. Ennui was weariness which had nothing to show: the tired hodman counted the courses in the wall. Languor pressed its nose against the pane, and dreamily questioned the vitality it mused on and envied. Earned leisure was most relished. Pure joy was a costly article. A little time for pleasure was precious; time for nothing else was burdensome.

The nervous tread of a true man.

Time was, even in Sub-Cœlum, when men generally were as indolent as they could afford to be. Unless compelled, they did little which was useful. Only now and then a high nature was created which worked from love, and was content with a tithe of the harvest. Nine parts to mankind was a generous division, and only a

INDOLENCE

great soul would spare so much. To such it was not sacrifice; his return was in multiplied blessings. Exemption from useful labor was the ambition or boast of nearly all. Trifling for selfish ends was therefore the business of most of those who could confine themselves to voluntary effort. They were perverted by a misuse of means. They relied upon the adventitious, till the natural, intrinsic resources denied them service. They went out of themselves for pleasure, and returned to find themselves empty. They built palaces, and existed in them the victims of ceremony and servants. They bought books to adorn libraries, which satirized them. They bought musical instruments as ambitious ornaments, and patronized the opera. They educated their daughters expensively, and saw them accept impertinence and imbecility for escorts and husbands. Their sons were indulged and pampered, till amusements were exhausted and occupation was purchased to keep them respectable. They rode in carriages so conspicuously elegant as to make them sacrifice comfort to propriety. Their horses represented so much capital that the weather and their health were consulted before

Trifling for selfish ends.

Sons indulged and pampered.

using them. Their acquaintances were esteemed for the rank they had and gave. Their houses were heated by furnaces to insure uniform temperature, and day and night they inhaled a baked atmosphere, and wondered at disturbed respiration. *Disturbed respiration.* Pipes conducted cold and warm water into chambers and kitchen, and they took poison in all that they drank and ate, and were surprised by palsy and an increase of nervous disorders. The wine-cellar, meant to be a depository of luxuries, became a resource against wasting vitality. The laugh of the fields and the streets was reproduced in ghastly caricature behind the parti-colored goblets. A joke upon the high price of bread redeemed a dullard, and the whole table from dullness. The children were cared for by nurses, and their natures modified by restraints and drugs, till feebleness and pitiful cries identified them. The doctor's visits were as indispensable as the baker's or hairdresser's, and the household ate as they dosed, by *Ate and dosed by prescription.* prescription. The priest dropped in to solace the moments between drugging and dressing. Life was taken up by the endless round of artificialities and their effects, till the struggles and wants of those they deplored compared with them as blessings. Their civilization at its worst, they slowly discovered that the inspiration of work was the spirit of life : that bread for the body, earned by exertion, was ambrosia for the soul. Sweet for the sweat it cost, it was sweeter for the promise it gave. It satisfied the appetite, but not the longing insatiable. The little feast was but a foretaste of fruition. The sickly atmosphere of affluence, tempered to tender throats and low enunciation, was gathered from cellars bordered by sewers, and choked a healthy nature, exhausted and exhaustive by exertion. The great lungs of outdoor labor inspired the upper air of heaven, and panted for inspirations from its source. To-morrow, on the way with the sun, would demand a full day's service, which to-day's fidelity must assure. To-morrow and to-morrow, and then the day supernal, long enough for any longing, an unending harvest and holiday. They realized that making money and earning it were different. Earning it was a reality ; making it a fiction. Money made money ; labor earned it. Bonds, proverbially, like infants, did best by sleeping ; labor was obliged to be awake, and faithful. A dollar, for a day in

Ambrosia for the soul.

Making and earning money.

the sun, was precious; a dollar, got in the dark, which could not be accounted for, was worse than want. Knotted hands told of the one; nimble fingers or nothing told of the other.

These views and activities developed manhood and personal freedom. Creature-comforts, more than were wholesome, were regarded with suspicion. In their simple philosophy, they were the lap of Delilah. They emasculated and smothered. Manliness, the thing every man should stand for, grew without them. Strong roots were made by strong winds. Careful culture and supports gave symmetry to the shrub in the conservatory, but the oak of commerce grew alone, amid storms. To the rude soil and the tempest it owed its texture, and it would bear the tests of the seas. They had seen how the branches of trees by the coast or on the mountain were sometimes forced by the merciless winds to grow one way; but the willful roots combined defiantly and forced themselves another. Character was so much resistance and endurance. They esteemed it a poor and disgraceful thing, not to be able to reply, with some degree of certainty, to the

Manhood and Personal Freedom.

Character.

simple questions, What will you be? What will you do? To cut the cable and launch away from conventional restraints and helps was the aspiration of every worthy man at some time in his life. His individuality felt fettered and shorn. Before he consented to surrender and be subordinate, he aspired to be tried by trusts, perils, and calamities. He had decided the fox lucky that left his tail in the trap. The muskrat, he had observed, would gnaw his third leg off to be free. Native manhood was shy of conventionality and patronage. It was inclined to be self-asserting, and was rarely arm-in-arm, but for recreation. It gave and took as it willed. It husbanded by determining without counsel. Its reserve conciliated what it would appropriate. It was democratic, essentially. It required and permitted alike. While it chose, it gave choice, without question. Freedom it claimed and allowed, an immunity without gyves. A receptacle, it could wait to receive, and would not obstruct nor be obstructed. A week was not idle that brought something, but a day was wasted if employed upon nothings. Its freedom was its strength, which modish subserviency acknowledged by obeisance. Its

The aspiration of every worthy man.

Immunity without gyves.

faculties were fitted for work by waiting for work worthy of them. Friction it liked, but not the attrition of mechanic movement. The principles it would freely use were as virginal and unhandled as when spoken of God. Ideal manhood stood for ideas, facts, and deeds. Rectitude identified it. The extrinsic was its foreground; the inherent its perspective, illimitable. Trials quickened and refined it. Wants supplied and pangs consoled it. Calamities became resources, treasures which did not waste, entailed for precious uses, perpetuated in goodness, or fame, or glory. In heroic days, plain food, in sufficient quantity, was all that was required. The appetite was kept whetted by labor, and digestion was as easy and unconscious as respiration. Sandwiches of corn-bread and bacon, with the fallen tree for a table, untouched and unpolished but by the winds of heaven, and the glittering axe for a platter, brighter than the brightest silver, made a delicious and brilliant dinner for the pioneer, after six honest hours of woodman's gymnastics. His simple and earnest life was ever a song or a prayer. The present was all thankfulness and the future all hope. His daily enjoyments, dearly

Rectitude identified it.

A delicious and brilliant dinner.

and honestly earned, were doubly blessed in health and sweet conscience by the Master Employer. His title to the acres he opened to the sun was directly from their Creator; and the bread they brought him was by the sweat of his own face. His future, in the steady serenity of heroic faith, appeared abounding in only such promises as his fidelity and devotion realized. His work and wants were so simple as ever to keep him close to the Giver. There was no middleman to divide his blessings or qualify his thanksgiving. His health the Helper, and his will the Assurance, his own short arm was long enough to reach the Bountiful and Everlasting.

Health and sweet conscience.

In the ordinary sense, the plebeian and the aristocrat did not exist in Sub-Cœlum. Society was so constituted, and men were so governed by exceptional conditions, that such distinctions were not recognized. Extremes met on the same plane. Personal freedom, self-respect, and the pride of manhood, placed men one with another. Every man a man, he naturally felt and acknowledged the manhood of every other. The uses of labor, of money, of intelligence, and of character, were held to be insepa-

The Plebeian and the Aristocrat.

rable. The responsibilities of wealth made the rich man grave, considerate, and modest. He felt his dependence the same as that of his less opulent neighbor. Frugality and liberality formed a just balance. Simple living and industry were resources to offset affluence. The same sum represented services recompensed and services rendered. Obligation and dependence were mutual. It was not for employer or employee to lord it over his copartner. In his freedom from the care of great property, the attentive citizen of moderate means esteemed himself fortunate as his eyes gradually opened to a knowledge of its perils and burdens. As he perceived the invisible hands reaching out from all round for the accumulated treasure — hands of mendicity and hands of cupidity — he better understood the delicate attitude of its possessor. The cares of honest poverty, he discovered, were not to be compared with the cares of hoarded riches. The piles of letters on the rich man's table every morning! The fulsome flatteries, ingenious and offensive! The threatenings, bold and insinuated! The schemers, soliciting money to balance prescience! Poor women, in the extremity of pride and

Mutual dependence.

The cares of hoarded riches.

distress, humiliatingly appealing for assistance! Reports of deficits in eleemosynary institutions! All to quicken sympathy and disturb the purse-strings. Agents were kept busy searching out the worthy. How could the rich man, with a heart in him, be free from anxiety and responsibility? His vessels were on the treacherous sea. His dividends had been lessened by a sweeping fire. His boy was a sorry expense. If he let his wealth accumulate, how was he to find secure and profitable investment?

To quicken sympathy and disturb the purse-strings.

THE VICES. The vices, in a great measure, had been eliminated, or had died out. Vast manufactories of drinks and superfluities had been abandoned. Tobacco was little used. Houses of sin were generally closed. Gambling was almost unknown. Occupations were numerously diminished. Those depending upon private vices almost ceased to exist. Horses were bred for moral qualities rather than for speed. The prize-ring was a thing of history. People wondered at its brutality as they read about it. That manhood should have been so perverted was one of the shocking things in their annals. As the ordinary uses of money di-

One of the shocking things.

The Change Revolutionary

minished, new employment was found for it. In proportion as the vices died out the virtues had been stimulated. The change had been revolutionary. Life was not the mercenary, sensual thing it had been. Chasing rolling bits of silver and gold had ceased to be its nearly universal employment. Pandering to extravagance and vice was no longer respectable. To elevate humanity, not to degrade it, had become the supreme object of civilization. Men became ashamed of what before they had been proud of. They studied, more and more, the laws of life, and the requisites to health and enjoyment. Expenditures being largely confined to comforts and necessities, not much money was indispensable. Hours of labor were reduced, and leisure was abundantly increased. Homes were supplied with every convenience, to make domestic occupations easy and attractive. The kitchen became a museum. Water, for culinary and drinking purposes, was perfectly filtered by simple and inexpensive means. Against flies, vermin, and insects of every sort, there was complete protection. The common rat and pestilent mouse had been so persistently, intelligently, and humanely pursued, that both

The virtues stimulated.

Hours of labor reduced.

species were nearly extinct. Nerves and sympathies being too precious to be wasted, heads of fowls were lopped off by ingeniously contrived guillotines. Simple and convenient apparatus for bathing was in every household. In the construction of commodes, of every variety and pattern, the utmost ingenuity was expended. Private offices, naturally disagreeable, were relieved of unpleasantness by attractive and luxurious appliances. Offal, fæces, waste of every kind, were consumed by fire, or reduced by chemical means to impalpable and scentless dust. The vices being no longer commodities, to any large extent, the multitudes dealing in them found other occupations. Genius was developed in unexpected abundance, and was felicitously applied, in innumerable ways, to make life abounding in comfort and happiness. Land increased in value as labor became more generally necessary to individual sustenance. The big diamonds and showy charms, no longer attractions in the gin-shops and brothels, were bartered for good acres and implements of husbandry. Dollars, got in the dark, were no longer many: all, with the few exceptions, were earned in the light, and under the sun;

Marginalia: Guillotines. Land increased in value.

and being limited to honest and clean uses, went a great way. Pecuniary independence was practicable and easy. A few hours each day supplied all that was requisite. Where wants were few and easily satisfied, is it any wonder the distinguishing names of plebeian and aristocrat were obsolete or inapplicable?

Increase of common sense and practical wisdom was a marked result of the new life. These high qualities appeared more conspicuously in all that they did. Their knowledge and experience were systematically applied. The comparatively poor, capable man, for that reason, became rich in resources. The economies and possibilities made him a master. How could he be utterly poor with unexhausted means — while anything remained to be done it was possible for him to do? His few acres produced marvelously. To the depth of three spades, sometimes, the light and gases were let in. Pulverization, fertilization, rotation, were matters of intelligent study and experiment, and there was certain increase in productiveness. Their kitchen gardens, more, even, than their farms, were attentively cultivated. A

Common Sense and Practical Wisdom.

Certain increase in productiveness.

small space seemed enough for a family. The vegetables were exaggerations, and their small fruits excelled in flavor and abundance. Cabbages and cauliflowers were favorites, and grew better by the affection bestowed upon them. Berries! — to know them you must taste them. Their flavor was an inspiration, and a joyful memory.

SMALL FARMS PREFERRED.
Farms were small in Sub-Cœlum, for reasons stated and inferred. Well tilled, they were found preferable to extensive plantations. Ploughing was deep. Drainage was complete. When necessary, irrigation was easy. Lakes on the mountains and high uplands, with perpetual streams flowing from them, supplied an abundance of water, and the topography of the country was generally such that the diversion of it from natural channels was not difficult nor expensive. Extraordinary care was taken in the selection of the seeds they planted. And they attentively studied the enemies of all kinds of grain and plants.

Entomology understood.
Entomology was so understood, that the habits of such worms and insects as they warred against were accurately known. How to exterminate them was always an

Knowledge Liberally Applied 81

interesting subject of conversation with agriculturists. The knowledge they displayed was acute and extensive, and was always liberally applied. Applied, mark you! for knowledge was not held of high estimation that was not practical and applicable. Do as you know, was an admonitory precept everywhere heard.

Fish-ponds were abundant, and great pride was felt in everything pertaining to piscatorial culture and art. The finest fish for the table, and the most beautiful for ornament, were always at hand. There seemed to be no limit to the supply. Ichthyological literature was exhausted to multiply them. Their nature was studied until it was understood. Just how to feed and treat them was known to perfection, and they grew in flavor and proportions accordingly. In the ponds, they were petted and caressed till they delighted in human companionship. They floated into your hand in a manner to invite sympathy and tenderness. Selection was made for the table with the least difficulty. In the streams, the varieties delighted in by sportsmen abounded. Every household had a cabinet of fishermen's supplies.

Fish-ponds abundant.

Selection easily made.

Nets, rods, hooks, flies — everything pertaining to the art — a veritable museum of utilities and curiosities. Everything was done to foster and elevate the art — nothing to disparage and degrade it. A great part of the poetry of life was inspired by the music of streams, and the skillful capture of their inhabitants. The man who did not delight in the temperate art of angling possessed no quality of the philosopher or poet. If he could not contemplate the running stream as an image of human life, and cast his hook into it as he cast his venture into the mysterious current of affairs, with only a hope or a guess of the result, he did not apprehend conditions. The shifting atoms, on their way to the sea, and the elusive fishes, are not more uncertain than the passing moments, and what they promise to us.

The poetry of life.

The cultivation of flowers was universal: every household had a garden of them. Bees, as a consequence, were generally kept and studied. Children, even, were wise about the wonderful creatures. Bees and bee-culture was a favorite topic of conversation. There was scarce any limit to the discoveries close observation had made

Bee-culture.

Talk of Bee-Keepers 83

of their habits and achievements. The talk of bee-keepers was as interesting as the talk of astronomers. It abounded in incidents and anecdotes worthy the attention of best-endowed minds. The ways of bees were as curious as those of men, and were freely used to illustrate human life and conduct. The philosophic uses of both, indeed, were interchangeable, without any great disadvantage to either. A knowledge of the wisdom of the little insects was not encouraging to the growth of conceit in the higher species. The more people knew of bees, the less self-flattering the estimates of themselves were. The parallels they constantly drew confused their notions of instinct and reason. Distinctions between them, and their limits, were never fixed, but constantly changing. No other creature under their care was so profoundly interesting. The suggestions of the apiary and its product were steady resources for mind and body. No food was considered quite so healthful, in certain conditions, as honey. The respiratory and pulmonary organs were helped by it, and its free use was regarded by many as a sure preventive of consumption. Well-defined cases of that dread disease did not

The ways of bees and men.

Honey healthful.

exist there, and the fact was accounted for in part by the general use of the sweet product. Oxymel had been an approved remedy time out of mind.

<small>PROPAGATION OF POULTRY.</small> Great attention was paid to the propagation of poultry. The barnyard was a picture. By careful selection and intelligent treatment remarkable results had been attained. Enemies had been destroyed or thwarted, and disease rarely showed itself. Eggs multiplied prodigiously. Artificial hatching was not in vogue. Too many of the fowls produced were deformed. Besides, in their nice sense, they did not like to disturb the course of instinct. Capons grew to great proportions and sweetness. The duck, in kitchen parlance, was all breast. The turkey increased in juiciness and flavor under improved feeding. But <small>*The bird of excellence.*</small> the royal peacock was the bird of excellence and preference. He adorned the farm and completed the banquet. His lofty, ostentatious mien made him an unfailing attraction. Guests at afternoon dinner-parties were entertained by his majestic strut and spread of tail and gorgeousness of color. An admiring word was enough to brighten and animate every fea-

The Royal Peacock

ther, and set him forth in all his glory. The gamut of ridiculous pride was in his dissonant notes. No other article of food commanded so high a price. In the poulterer's stall he was adorned with ribbons. Just the time required to ripen him perfectly, was a question gastronomers were ever discussing; and how most divinely to cook him, was a subject that inspired genius. Poets sang the royal bird, and painters exhausted their pigments to imitate his tints. Unique ceremonies were performed over him, as he lay in his fragrance and juiciness, on the banqueting table, before anatomy divided his bones, and laid bare the depths of his bounteous bosom. The skilled carver, as he cut away the succulent flakes, was expected deftly to show them in such light as would display their translucency and lustre. Times when the peacock was the special gastronomic glory, were occasions of faithful and triumphant record. Draughts were made of the table, and the names of honored guests were appropriately set down in roseate colors.

Adorned with ribbons.

The Sub-Cœlum oyster was the best of all the sixty or more known species. The

The Sub-Cœlum Oyster.

Sub-Ca'lum

Favorable flats for transplanting.

beds on all the shores were extensive and abundant — especially at the mouths of the great rivers. Favorable flats for transplanting were at convenient distances from the great beds. The greatest were in shallow water, not much above a dozen feet in depth, making the dredging process comparatively easy. Transplanted to the marshes, fed perpetually by innumerable rills of sweet water from the mountains and highlands, flowing through beds of odoriferous herbage, imparted a matchless flavor to the universally beloved mollusk. The bays had been stocked till the multiplication was incalculable. Industry and science had done wonders. The delicious bivalve was of unlimited consumption, and cheap. Raw and cooked, he was served in every attractive manner. Only the perfectly healthy oyster was marketable. The slightest show of disease consigned him to the basket, to be fed to the poultry and the

The wading oyster-catcher.

fishes. The wading oyster-catcher was hunted industriously, and did not multiply. In very many cases the peculiar bird was made to lose his predatory habits by domestication. Thus diverted in nature, he formed a handsome addition to the park and poultry yard. Once every year, a day was set apart to the celebration of the oyster; and oyster-holiday was joyfully welcomed and universally kept. Public tables groaned, as we say, with the incomparable marine production. It was the festival of the people. They met together as one great family, and transfused a spirit of love and patriotism from one to another. If any estrangement existed between friends or neighbors, it was expected to end with that day. New acquaintances were formed, and a flow of new blood fused society to a higher healthfulness. Prepossessions and jealousies and envies vanished from sound hearts. Grudges were never more than a year old. Sullen malice or malevolence, of longer existence, was treated as disease, or occasioned unenviable distinction. Socially, an invisible guard was set round it, as around a dangerous malady. The moral indebtedness of the population to the annual festival was incomputable.

Public tables groaned.

The grape, of different species, and of many varieties, had been indigenous from the beginning. Soil and climate were adapted to its growth. In the wild state, the fruit was inviting and palatable, but under intelligent cultivation it was unsur-

GRAPES AND WINE.

Vineyards everywhere.

passed. The hills everywhere were adorned with vineyards. Old and young found congenial employment in them. Favorable conditions made it possible, without great artificial aid, to have the best varieties the year round. Kinds best adapted to the table were cultivated to be almost seedless. Grapes were so abundant that they were very cheap. All enjoyed them without stint. Wine-making was one of the active pursuits of the country, and those engaged in it were proud of it: the cleanly vats and the delicate manner in which the clusters were trodden, gave proof of it. Splashed ankles of fair women added picturesqueness. The red and purple upon lustrous semi-pellucid extremities were tints to be remembered evermore. Artists and bards made the most of them. Attempt was made to employ young elephants to press out the juices; but the innovation was discouraged. Opportunities of fair and just rivalry were not to be restricted. Endowments of nature were not to be thus disparaged. Everybody drank wine, as he did water, or milk, for refreshment and nourishment. Nobody thought of questioning the morality of its use. It was upon every table at every meal. As great

Splashed ankles.

Everybody drank wine.

A Man Drunk was Odious 89

pains were taken to keep it pure, it was found to be healthful. Drunkenness from wine-drinking was unknown. It was from distillation that the mischief came. Fortunately, the strong liquor was little used. Public opinion was against it. Reputation was affected by its free use. Drunkenness was treated as disease. Victims of it were separated from the general public. A man drunk was odious. If shame did not prevent a repetition of his offense, he was in danger of being considered incorrigible, and of being treated as such. Examples of the kind helped to educate the people to right conduct. They did more to instruct than all the didactic poems, essays, and addresses. Their effects were thorough, and went to the sources of the evil. Societies were not formed to exterminate the drunkard, nor to make a pet of him. He was held responsible till officially declared otherwise. Drunkenness was attacked as a moral disease, not to be cured by salves nor embrocations. The miserable habit would die out when better standards and inclinations were established. The sin was personal, and not of society. The comfort and innocent pleasures of the many were not to be restricted by the

Drunkenness from wine-drinking unknown.

Examples educated the people.

The sin personal.

excesses of the few. The mode of reformation was not by absolute self-denial nor prohibition. The joy of the common heart was not systematically restrained nor repressed by individual instances of voluntary excess. The good things of Sub-Cœlum were to be enjoyed, and not to be abused. Good wine was inseparable from the life, which comprehended all that was excellent, and a just and generous enjoyment of it. To rejoice was better than to groan. Ills were forgotten in good-fellowship. Misery was not helped by lamentation. Dolor was no cure.

The joy of the common heart not restrained.

ENDLESS ORCHARDS. Fruit trees were planted at each side of all the public roads. Not so near together as to impoverish or seriously shade the land contiguous. This utilization of the public spaces supplied the choicest fruit in abundance to everybody. All any one had to do was to gather it; but it was a grave offense to damage in the least the trees. The laws regulating this wise provision were of the strictest character, and were rigorously enforced. Public opinion, however, was a better protection than any enactment. The people were proud of their endless orchards, as they called them, and

The people proud of them.

Belonging to Everybody

guarded them with scrupulously jealous care. It was the rarest thing that a tree suffered from ill-usage. The Commonwealth planted the trees and maintained them. The old, or sickly, or ill-bearing were from time to time cut down, and young, vigorous, promising ones put in their places. The long lines of thrifty trees were a delight to see. In bloom, they filled the imagination. The bees made them musical. Filled with luscious fruit, they stimulated the palate, and made happy the birds. Such walks and drives, bordered by fragrance and richness! Belonging to nobody, but to everybody! In full fruitage, the bounty was in fruition. The Government, if a sentient, sentimental thing, might have realized the blessing, and led in the thanksgiving. Patriotism, under such conditions, was as natural as filial affection. Incivism was not conceivable. Generosity, too, was spontaneous. Easy supply was inseparable from free giving. The common heart was not circumspect nor prudential. The humanities quickened it, and made it unconscious in all good offices. Better men and women were but the natural result of the never-ending munificence.

Ill-usage the rarest thing.

Fragrance and richness.

Generosity spontaneous.

HIGHWAYS IDEAL.   Their highways were ideal in excellence. They were made of the good materials supplied by their valleys and mountains, and were as level as practicable, and perfectly drained. Grades were mathematical and easy. Impediment of any sort was not permitted. A single draft-horse would draw as great a burden as the most substantial of their wagons would bear. It was a joy to ride or drive on their roads; and the horses felt the inspiration. Vehicles, almost self-moving, were in general use. Everybody had some independent means of mechanical locomotion. Chariots large and chariots diminutive, with sails, with batteries, with wings, glided along without equine assistance. Happy children! happy women! happy men! Under the blue dome was ever anything more joyous?

HOW CITIES AND VILLAGES WERE LAID OUT.   Their cities, towns, and villages were laid out in squares, with streets running, as we say, from northeast to southwest, and from southeast to northwest. Laid out in that manner, neither side of any street had any advantage. Sunshine and shade were the same on both sides. Property, in consequence, was alike desirable

Sunshine at a Premium 93

on either side, and, other things being equal, commanded the same price. Sunshine being at a premium, everybody wanted all he could get of it. Where houses were separated sufficiently, the sun shone on every side alike. Every outside room had the sun a part of each day. Windows, as a rule, extended from floor to ceiling, and the air inside was sun-swept and purified diurnally. In chambers, beds were drawn out to receive the sunshine in floods. Musty and damp beds were unknown, as were certain diseases that breed in perpetual humidity and shadow. Free sunshine and free air were in permanent fashion, and were not intercepted nor excluded except when necessary. Perfect ventilation was a desideratum, and was attained as nearly as possible. The sweet air! Had God Almighty intended they should stint themselves in it, would He have poured it out all round the earth forty miles deep? Sun-painted complexions were preferred. Paleness was deplored. The pride of the women especially was their high health and high color, which they attributed largely to unlimited light and pure atmosphere. Living much out of doors, they unconsciously caught the freedom of the

Sun-swept air.

Sun-painted complexions.

elements. Their eyes were strengthened and brightened by being accustomed to great range of vision. One of the pastimes was to count the birds, or other small objects, so far away as scarcely to be seen. Every considerable residence was provided with a room lit only from above. The purest glass was used, and the moving clouds were as visible as from out-doors. Convalescents and invalids rejoiced in the pure light and living frescoes. On cloudy days and moonlight nights the sky-lit rooms were most attractive. A day spent in one of them was like a day spent in another zone.

One of the pastimes.

DRAINAGE. Drainage was as carefully considered as air and sunshine. In the location and construction of every house, provision was made to get rid of every drop of surface water not purposely caught and appropriated. Effects of neglecting thorough drainage appeared in familiar statistics. In old maps they pointed out the routes by which epidemics had traveled, invariably over spaces imperfectly drained. Filled-up marshes, and little streams leading to and from them, had been the abode of wasting and rotting diseases, before the houses

Typhus had no Chance 95

that covered them had been pulled down, and the land thoroughly drained, according to scientific system. Cellars and sewers were rigidly inspected. Typhus had no chance to burrow or linger. Rich people had no advantage over their less fortunate fellow-citizens. The provision was general, and people of limited means and the opulent were alike rigorously governed in every detail pertaining to the public health. Humble abodes were not more frequently visited by disease than palaces, and there was not an unhealthy locality in any town or city.

Rich people had no advantage.

Light and heat were obtained almost entirely from water. After long-continued experiment, the elements to produce them had been separated and applied. Every house was illuminated and warmed at a moderate cost. The streets of cities and towns were brilliantly lighted. The process was ingenious, but not complicated nor dangerous. Besides being simple and cheap, it was easily manageable. Temperature was self-regulated. All you had to do was to determine the standard, and the machinery did the rest, without considerable variation. Cleanly, too, the system of light-

LIGHT AND HEAT.

ing and heating was, without measure. Housekeepers were not troubled with dust, nor smoke, nor vapor. With the perfect system of ventilation, the air was kept pure without difficulty. Nerves and brain were stimulated by it, and the lungs delighted to take it in generously. It was the general belief that mind and body were both helped by the improved method of heating, and great hopes of increased intellectual and moral development were fostered by it. Exalting tonics and enrapturing odors were diffused through the atmosphere at pleasure. Talent expended itself in producing essences and tinctures and stimulants of paradisaic delicacy to be so employed. On great occasions the light produced rivaled that of the sun. The whole atmosphere seemed to be aflame. The effect was magical. The smallest thing was made visible, and all things were beautified in appearance. Men appeared more manly and women more lovely. The pretty children seemed just to have descended.

The air was kept pure.

Public edifices were not built to endure forever. Substantial enough and suitably adorned, they were meant only for a gener-

PUBLIC EDIFICES.

Temples of Justice

ation. Instead of expending a million in constructing one of their temples of justice, to stand for a century or two, one fourth of that sum was found sufficient to erect a suitable structure, to last for an age. *To last for an age.* Thirty years' time was found to be about the limit of a decent degree of cleanliness and purity for a public building. The foul gases and scents and creatures would get in, and no amount of precaution or care would keep them out. It was discovered that the only way to destroy them completely was to take down the building. The structure to succeed it was built after the latest models, and was adapted to the generation that was to use it. Better drainage was had, and better provision was made for ventilation and lighting. In every way the new building was an improvement on the old, and was better adapted to the purposes for which it was intended. This habit of general demolition and reconstruc- *Demolition and reconstruction.* tion was for economic as well as sanitary reasons. Experience had proven that repairs alone, on a million structure, to say nothing of the item of interest, exceeded the cost of new buildings. Experiments of architects and plumbers were not made except at great expense, and as often damage resulted from them as benefit. At best, modification and adaptation made it an old building. While the architecture changed with each new edifice, much care was taken to limit the cost of it. Showy ornamentation was strictly avoided, as not in agreement with the public taste or public policy. Newness and freshness were preferred to decay and dinginess. Distaste for soiled finery was pervading — it extended even to neglected ostentatious buildings. Architecture, therefore, looked to simplicity and cheerfulness, and scrupulously avoided whatever might appear sombre or involved. The public was generous to the limit of reason, — extravagance they did not permit. Expenditures must be prudent and exemplary. The citizen was not to see in the public what would be condemned in himself. The universally adopted code of morals forbade the expenditure of public money without necessity, or beyond what was reasonable or proper. Reckless dissipation of the people's money was of rarest occurrence.

Economy consulted.

HOTELS. Hotels, also, for the public entertainment, were built to last only for a generation. Experience had taught that, in spite

The Old-Hotel Smell

of all the soap and paint and disinfectants that could be used, they would grow offensive to the olfactories. The old-hotel smell was pronounced the most objectionable and noxious of all the variety known to the nose of man. It was the product of cellars, sewers, closets, et cætera, and contained a portion of all the subtle poisons known and unknown to chemistry. Only the sunshine and fresh air would dissipate it. Proverbially, the newest hotel was the best. The public, as a rule, systematically passed by hostelries where for many years human beings had eaten and slept and performed every private office. Pollution bred there. *Objectionable and noxious.*

Not a bell was heard from any building in Sub-Cœlum. Years and years had elapsed since bells had been used to call the people together for any purpose. Everybody had a clock in his house or a chronometer in his pocket, and bells were not regarded as necessary. Besides, the noise had become generally distasteful, and the common feeling and the common sense had prohibited it. After every attempt had been made to improve them in tone, it was decided that the best results could hardly *Bells.*

be called musical. The most complete chimes, in the common ear, were little more than discords — consonance or harmony was not in them. The highest excellence in music having been attained, the public ear was so acute as to be shocked by mere noise. Every one having a taste for the divine art was encouraged to cultivate it. Scientific training had made the majority pretty good musicians. Mere noise, to the extremity of possibility, was avoided. Exquisite and exalted strains precluded it — even the consciousness of it. Absolute softness and sweetness were desiderata. The tones of forty instruments were so perfectly blended that you hardly heard them a few rods away. But bells had been abolished for better reasons. The people had increased in thoughtfulness, refinement, and good-breeding, until they would not permit what might be regarded by any considerable number of persons as unnecessary disturbance. Sick people, people in distress, were thought of in all that pertained to their comfort and protection. Jingling, jangling, tintinnabulary noises, to rend sensitive nerves and hammer inflamed brains, were tortures to the unfortunate that considerate civiliza-

Shocked by mere noise.

Considerate of sick people.

tion did not tolerate. People who from any cause needed sleep were remembered and protected. The voice of one, in extremity, was heard and heeded by the multitude. Majorities were considerate of minorities. Might did not make right. The pervading thoughtfulness of others was one of the distinguishing charms of the population. It quickened perception of justice, and tenderly regarded weakness. It made aggressiveness offensive. Hardness was barbarity. Noises, irritating to many, and not necessary to any, like those produced by loud-sounding bells, were dispensed with, as not in agreement with their philosophy of life. Their scheme of civilization was to make everybody happy — nobody miserable.

Might did not make right.

Music was so generally cultivated and enjoyed that it largely governed the life. It was vocation and avocation — employment and diversion for mind, body, and spirit. Taste and ability for it had come down through the generations. It seemed as natural to them as any appetite, and as necessary as to breathe. They could, most of them, sit down at an instrument and practice for hours together without weari-

MUSIC.

ness or nervous disturbance. If ill effects followed application, continuance was discouraged. The pupil was not thought suited to the art to whom it was labor to study and practice it. To force him was considered detrimental to health and happiness. The result was, that while everybody enjoyed music, not everybody continually attempted to produce it. The population good-naturedly put up with tyros, at the same time they took pains to protect themselves against them. Isolated halls were provided for students to practice in. Anybody could not blow his horn anywhere without authority. Brass-bands, except the few that were distinguished, were permitted to play in the streets and public squares only on certain holidays. At other times they were officially relegated to the fields and forests—to play only in the daytime. The night was held sacred to silence and sweet concord. Learners in households were only heard at certain hours in the morning, when ears and nerves were most enduring. The general musical taste and education of the people did not tend to unfit them for other occupations and avocations. It was possible for a performer or vocalist to get a liv-

A good result.

The night sacred.

ing by other means, however proficient he might be in his art. Idlers did not abound in consequence of the prevailing passion and acquisition. Musical societies of every character were permanently organized — small for private enjoyment, large for public exhibition. The home entertainments, in which music predominated, were superior. Imitations of sounds of every sort were produced by the voice and by instruments. The Æolian harp itself was imitated, as well as the notes and cries of birds and animals. The moan of the sea and the murmur of the brook were reproduced with surprising exactness. The birds in the cages joined in the concert. The cock in the barnyard responded to his own notes. Fun and enthusiasm mingled. But in the music of Heaven — the oratorio — they were happiest and most transcendent. The sublime choruses kindled the imagination and enraptured the soul. Not a thought of noise was suggested or impressed. Discord was not. Harmony prevailed, and governed, to the last degree. You left the great auditorium full to the throat, and the eyes, of the glories that are, and the glories that are to be, evermore.

Musical societies.

The music of Heaven.

Sub-Cœlum

POETS AND POETRY.

Great poets there were in Sub-Cœlum; but not many. Their names being short, you could utter them all without taking a breath. Poetasters, however, were numerous; and rhymsters without end. Verse-making was one of the common amusements of the people. Much of their correspondence was in verse. Facility in the use of language, and their musical sense, made the process easy. Rhythm and rhyme were one to them. But poetry was another thing, and attempts at it were not received with favor. High standards made it unattainable by mere labor.

The Maker made the poet.

The Maker made the poet. The poetical view of nature and man they regarded as the clearest view, agreeing with one of the great sages, that the meaning of song goes deep. The poet was, to them, indeed a seer, a prophet, a soul divinely inspired. From him more even than from the priest, they had evidence that Sub-Cœlum was overspanned by a veritable though invisible Super-Cœlum, city of the Eternal God. Therefore they held no such foolish saying as that a proposition has in it more truth than poetry, for poetry to their apprehension was the nearest approximation to absolute truth that human language could achieve.

Poetry to their apprehension.

The Art of Poetry

To say that a statement was true as poetry, was to exhaust the power of exact speech. The person in their community who had no sense of beauty, no ear for music, and no susceptibility to poetic influences, was looked upon with pity, much as in other parts of the world humane people regard an amiable and intelligent dumb animal. For the Sub-Cœlumites were the most tolerant and forbearing of mortals, largely because they were suffused with the sweet light of the imagination. They could even bear to have fellowship with men and women who were destitute of humor, that most celestial virtue. To them Poetry and Humor were the nectar and ambrosia of the gods. In their palaces hung the portraits of all the great Makers from Homer and Æschylus to the nobler bards of their own realm and time. Boys and girls were brought up to honor the name of Poet, and to fashion their lives according to the supreme morality of the immortal poems which interpret both human truth and divine revelation. The art of poetry itself took on a wonderful and almost incredible development under the new conditions of life and new motives to action existing in their civilization. Like every other expression

The most tolerant and forbearing of mortals.

Brought up to honor the name of Poet.

of man's consciousness, in that extraordinary country, poetry was large and free, and adequate to nature. The sublime and beautiful forms which verse assumed were innumerable. There was a general breaking loose from conventional fetters, — an infinite expansion of the laws of rhythm, melody, metre, stanza, and trope; — the inspired soul of the creative genius put on robes of singing splendor, and revealed the infinite Love and Beauty and Power through the medium of words. All the people studied and practiced, in some degree, the science and art of poetical composition, as they did the elements of music, not for the purpose of setting up as poets or musicians, but in order to be able to appreciate and enjoy the superb productions of the mighty masters.

A general breaking loose from conventional fetters.

MUSICAL VOICES.

One of the most interesting results of their temperate and cultivated life was the great proportion of finely modulated voices. Very many of them were extremely musical. Voices hard, harsh, husky, disagreeable, were exceptional. Tones, as a rule, accorded with habits, dispositions, and acquirements. Free, almost entirely, from excesses of any sort, kindly in nature, and

Slow and Deep Breathing

thoughtfully intelligent, gentleness and sweetness of expression were only natural to them. Vices and violences had not disordered their speech. Gluttony and drunkenness had not inflamed the membranes. Breathing was free and unconscious, and was little more rapid when awake than when sleeping. Slow and deep breathing had long been practiced advantageously. A dozen inspirations to the minute were not very uncommon. Increased strength, flexibility, and richness were added to the voice by the good habit. Pretty long sentences were easily and naturally spoken without taking a breath. Lost or artificial teeth did not affect their articulation. The insides of their mouths were not covered over with gold or other substance to abrade or indurate the delicate surfaces, and consciously modify expression. The dress of men and women did not interfere with the natural growth and expansion of their chests. It was the rarest thing that breasts were not broad and arched. Throats, too, were round and full, from never having been compressed or hurt by vicious dressing. It was considered an outrage upon nature to do anything that would interrupt in the least the free growth

Breathing free and unconscious.

Throats round and full.

of any part of the body — especially of the life-giving apparatus of respiration and enunciation; on the contrary, everything was done to promote its completest natural development. The slightest disturbance of its functions was anxiously observed, and corrected, if possible, as interfering not only with individual comfort, but the general happiness. Inhaling tubes were freely used, to make slow and deep breathing habitual. Men and women walked miles at a time breathing entirely through them. Instances were not uncommon where the circumference of the chest had been increased from one to two inches in a year by frequently using them, and that without increase of bodily weight. Great inspirations of oxygen moistened the spine and beaded the brow, and prepared them for any intellectual or moral work, better than by other possible means of stimulation. But the melody of their voices was far from being wholly owing to their physical life and training; their high moral natures and cultivated intellects contributed as much or more to produce it, by reacting on their sympathetic bodies. Conversation between highly enlightened and humane men and women, upon worthy subjects, was

The life-giving apparatus.

Great inspirations of oxygen.

Conversation.

Their Good Readers

charming indeed. Tones were as varied as the notes of the harp when played upon by the winds. Thought and feeling, in gradation and development, were unconsciously betrayed in ever-varying modulations. Voices flowed, like the full-running brook — now slow, now rapid; rippling joyously; then descending, where it was still and deep, to swell again in fuller richness, with the glow of imagination and sentiment. To hear one of their good readers read was a very high order of entertainment. Not an affectation or trick of the self-conscious elocutionist was visible in the exercise. He lost himself in the printed page, and his voice echoed its thought and emotion. The conversation between the Twa Dogs appeared the most natural of dialogues. The interlocutors seemed indeed men, until the invincible humor compelled you to remember they were only dogs. The battles in Homer were as real as any conflicts could be. In passages of Job and Habakkuk you felt in full force the sublimity of supra-imagination. In the scene of the White Rose in the Paradise of Dante you had a vision of highest heaven. A lofty meaning was revealed that might have astonished the poet himself.

Thought and feeling.

Lost himself in the printed page.

Scene of the white rose.

TIGHT DRESSING.

Proportions must be artistic.

Fashion exacting.

Tight dressing was not fashionable in Sub-Cœlum. The people were proud of their natural bodies and sound children. It had been a great while since any general effort had been made to divert or thwart nature. Occasional attempts in that way were always attended and followed by the same results. Time was when distortion of the body was common. It was thought beautiful to be out of nature. The shape was fixed by the artist's patterns, no matter at what cost of pain or violence. The rules of tape and scissors were remorseless. Proportions must be artistic. Form must be fitted to the mould. Life was absorbingly artificial. Balls and calls and parties and operas and shopping left little time for anything else. Children were an incumbrance. Nurses, most trustworthy, might be obtained; but the mother could not withdraw her mind wholly from her offspring. The success of her friend's magnificent entertainment would be qualified or marred by her uneasiness and anxiety. The tastes and requirements of gayety and maternity were incongruous. Fashion was exacting, and would not let her votaries divide or suspend their worship. Out of fashion, out

The Race Threatened

of the world, was one of her maxims. Out of sight, out of mind, was another. Lists of friends were continually being revised, and a chance would occur of being left off. Babies, she said, were vulgar; *Babies.* they were troublesome and spoiled the shape. Her rule was omnipotent while it lasted. Only the general decline of health and weakness of progeny abated her power. The vigor and happiness of the race were threatened, even its existence. Nervous disorders multiplied. Soundness of mind as well as of body was slowly sapped. Three or four successive generations showed marked declension and degeneracy. Society, only after such convincing results, became alarmed at her follies, and set about righting herself. Revolution was pretty nearly complete. It became fashionable to keep good hours, to eat healthful food, to wear loose, comfortable clothing, and to carefully avoid *Nature not* any interference with nature. The beauty *to be interfered with.* of the race — of men and of women — increased. They were healthier and happier. They enjoyed, more and more, their homes and children. Gayety abounded of the natural kind. The joys of life were the joys of health. The Style and Mrs.

Grundy did no longer govern absolutely. In fact, it became the fashion to be healthful and natural and robust. Good complexions came of right living. Paleness or sallowness was exceptional. Unconsciously elastic bodies and sound minds predominated. The young, left to nature, were as free-bodied as young animals always are. Domestic life was ideal. The atmosphere of well-ordered homes was the best under heaven.

Complexions.

It was deemed the greatest part of their felicity to be well-born — of parents with sound bodies, sound minds, and correct principles, and to inherit the same. It was asserted that no one ever changed his character from the time he was two years old; nay, from the time he was two hours old. That he might, with instruction and opportunity, mend his manners, or alter them for the worse, as the flesh or fortune served; but the character, the internal, original bias, remained always the same, true to itself to the very last, feeling the ruling passion strong in death. They believed, with the same authority, that the color of their lives was woven into the fatal thread at their births; that their original

A Felicity to be Well-Born.

The fatal thread.

Each Man's Destiny

sins and redeeming graces were infused into them; nor was the bond that confirmed their destiny ever canceled. It was said, and believed too, that, by whatever name you call it, the unconscious was found controlling each man's destiny without, or in defiance of, his will. Also, that all individuals were the outcome of past influences. Generations lived and thought and acted that each one might be what he was. Were any link in the chain of heredity lacking, he would be different in aptitude, in capacity, in very form and appearance. The absence of some faculty, the feebleness of some disposition in some one or other of his ancestors, were sufficient to vary the results in his own person. Ah! they thoughtfully and sadly exclaimed, if only full-grown men and full-grown women, with sound bodies and sound minds, were permitted to marry! Conscience, integrity, and reason, as far as possible, were educated to that end.

The bond never canceled

The chain of heredity.

The population was of many races compounded. The blood of many peoples had been infused into it. So composite in its character, social problems had been slow of solution. Prejudices of race had been

A COMPOSITE POPULATION.

a great hindrance. The more refined and gentle had been shy of the rude and aggressive. Conservatism had resisted the clamors of new blood. Power grew timid from variance of interests and suscepti- *Sharp antagonisms.* bility of change. Sharp antagonisms kept society continually at the point of boiling. Good had come of all this clashing and fermentation; but the people wearied of it. Reaction was inevitable. It came; and with it a disposition to liberality. Fusion seemed not so difficult. Opponents cooled, or went arm-in-arm. Individuals graciously coöperated for the public weal. Notions gave place to opinion, and opinion to reasonable judgment. Where clamor had been bedlam, deliberation reigned. Like a mighty stream of many tributaries, progress was no longer checked and fretted by obstructing jealousies and hatreds. Minor differences, in thought and in action, *Race prejudices.* were tolerated. Race prejudices gradually gave way, and bigotries. Fibres intermingled and blood interfused. Distinctions were obliterated by intermarriage. Freedom of taste was indulged. So many varieties, the faculty of discerning enjoyed great scope. Each race had supplied its characteristics, physical, intellectual, and

Vigorous Men and Women 115

moral. Temperaments, from the frigid to the fiery, were in contact. Every color of hair and almost every tint of complexion. Voices coarse, and musical as Apollo's lute. Noses straight, aquiline, and snub. Ears delicately transparent and ears rudely drooping. Lips refined and lips voluptuous. Deep chests and shallow, with great lungs and feeble. Muscles of ropes and apologies for muscles. Alexanders tall and Thumbs diminutive. Bearded and beardless. Every variety of man and woman to select from. Marriage was not interfered with, except in cases of close relationship. Complexions, as a result, were often very striking and beautiful, and figures produced of remarkable mould. Vigorous men and women were the rule. The exceptionally puny of both sexes, kept apart, not considering themselves proper subjects for wedlock. The population steadily improved in every respect. Intellect was quickened and the heart softened. Temperament, especially, was refreshed and stimulated. Emotion was indulged; feeling was exhibited without exciting derision. Children were born happy, and were not regretted. Grace was in their attitudes and music in their voices. Na-

Every tint of complexion.

The population steadily improved.

ture had free sway. Aptitudes developed early. Inherited traits were conspicuous. It was soon perceived what the child desired, and was born to do, and he was educated and encouraged accordingly. It was a maxim of one of their sages, and they acted upon it: Of that which a man desires in his youth, of that he shall have in age as much as he will. Elements of power and culture were realized in consequence. Love of thought and love for the beautiful appeared spontaneous and uppermost. The man or woman was what nature meant him or her to be. Old family portraits showed many shades of complexion and great variety of conformation. Extremes met in every collection. Faces so dark as to require light backgrounds to make them distinctly visible were close beside others, delicate, fair, and rosy. Rudeness and coarseness contrasted with high-breeding and refinement. Looking at the differing portraits, it was not difficult to account for their liberal and enlightened civilization. Nature, in a fateful, mysterious way, had propitiously brought about the inevitable. Toleration and upward growth were necessities. They must respect each other, and be better.

A maxim of one of their sages.

Contrasts.

Joys of Wedlock

Weddings in Sub-Cœlum were strictly private and unostentatious. Not that marriage was more uncertain there than in any other part of the universe. It was a test of character, the result of which was everywhere and always past anticipating. The least promising often turned out the best, was a proverb. The miseries of wedlock, they said, were to be numbered among those evils which cannot be prevented, and must only be endured with patience and palliated with judgment. Its joys were the greatest known to mankind — inestimable and inexhaustible. The dream of hope and expectation, when realized, was the one incomparable and never-ending felicity. The worse than blanks with the prizes made the drawing always dangerous, and it was deemed prudent to postpone the celebration till a year or two after the wedding. These occasions of rejoicing were frequent, and were participated in heartily by friends and relations. Fate and fortune had been bounteous, and thanksgiving was spontaneous. Fact was commemorated, not hope celebrated; happiness was realized, — better than all anticipation. Man and wife were congratulated, not bride and bridegroom. Whatever of fret and irritation

WEDDINGS IN SUB-CŒLUM.

The dream of hope and expectation.

Man and wife congratulated.

had been experienced, the calm had come, and the open sea, with a bright sky over all. Ideal was real. Misconception had given way, and each appeared better to the other, though different. They understood each other, and were incorporated. A child perhaps had blessed the union, and the household was a home, in all that the word implied. Presents were simple and appropriate — useful and to be used — and were not in any sense satirizing or vainly showy. A different moral atmosphere pervaded one of these commemorations than that of a bridal celebration. At the wedding, mystery and uncertainty made the thoughtful grave ; only the giddy were unqualifiedly joyous. Shadows and clouds did not appear to their hopeful eyes. Plain sailing only was thought of, without variable or conflicting winds. Compounding incompatibles had not entered into their intellectual chemistry. Fusing dissimilar natures they had not thought of as one of the difficult things under the sun. Love, the inspiring amalgam in their theory of life, would as often fail as succeed in the conflict of diversities. Interest and necessity and pride did not enter into their calculations of connubial existence. They did not

Ideal was real.

The inspiring amalgam.

The Omnipotence of Silence

calculate at all; they only dreamed. Concession, compromise, surrender, they did not see as necessities. The omnipotence of silence, in extremity, was not comprehended. Wise Sub-Cœlumites, to celebrate marriage a year or two after the wedding ceremony!

At one time six unmarried persons, three of each sex — guests at one of their unpretending watering-places — were in a sail-boat together, becalmed. For entertainment, it was determined that each one should tell the rest, in a word, why he or she had remained single. Acquaintances but for a week, and not likely ever to meet again after a fortnight, they spoke with unqualified frankness. Of uncertain age, they were not without experience. *Reasons for remaining single.*

The first to speak was a gentleman, say of forty-five or fifty years. The governing reason, he said, why he had not married, was self-distrust. Early experience had taught him the inconvenience, if not the distresses, of poverty. He remembered the sacrifices of his mother, and had resolved that his wife, if fated to have one, should not be subjected to like expedients and hardships. At twenty he was enam- *The first to speak.*

ored of a fair girl — the fairest, by far, he ever had seen. She filled his eye, his mind's eye, his imagination. She was very lovely. He was shy of her presence, but he could not keep entirely away from her; *She fascinated him.* she fascinated him completely. He had the will of a full-grown man, with a few years of initiatory experience in a respectable occupation; but all, indeed, of real life, was yet before him. He did not know the stuff of manhood that was in him: he had not been measured and tried by affairs. His intellectual and moral grappling-irons might be unequal to the grasp that was necessary even to ordinary success. He dared not meet the incomparable girl alone — he was sure to tell her he loved her if he did. There was not a word or a caress that all the world might not have heard or seen. He subjected himself to severest self-questioning. If he asked her and she said yes, what was he to do with her? Over and over he turned the problem in his mind, through anxious days and sleepless nights. Not without many a struggle he distrustingly determined that he had no right to ask her — the all-worthy incarnation of super-excellence — to take the chances of life with him. Heaven sent

Violence of Disposition

her a more courageous lover, and she died an idolized wife and mother. He might say he had prospered in the world; but he had never met with another who was the same in his eyes and affections. And was it possible he could love one inferior to her?

The next gentleman to speak was younger by a few years. He had a devil of a temper, he said, and all of his life he had been afraid of its consequences. Quick as a flash, he had once thrown a hatchet at a boy for a slight indignity. Placid as he appeared, the violence of his nature could not be comprehended. With plenty of red in his complexion ordinarily, in a rage he turned white as a sheet. In one of his fits he dared not look at himself in the glass. At such times a vicious grandfather looked out of his eyes. The dangerous old man was a terror as long as he lived. Two or three times he had been locked up as insane. He himself was in constant dread of the same treatment. He did his utmost to govern himself; but once in a while, in spite of all that he could do, the Satanic in him would break loose. His acquaintances were chosen for their forbearance and placidity. He had an eye

The next gentleman.

In constant dread.

to the same traits in his employees, and paid a premium for them. Once, a conflict with one of his workmen nearly cost him his life. He had also exposed his evil disposition in a court of justice, while giving his testimony. Through the good influence of his mother he became a member of a church society; but his dread of becoming a disturbing element made him withdraw from it. His best reliance as a safeguard was his ability to control a strong appetite for drink. The possibilities of his evil nature were terrifying enough without artificial stimulation. Think of it! A man with such tendencies to marry! God help the poor woman who risked a union with him! The novel irritations of the relation would have been sure to develop the bad in him preternaturally. The tiger and serpent might never be wholly quiescent or torpid. One experience of the tender passion, he said, was enough. His sweetheart had knowledge of his success in the world, and seemed disposed to encourage his suit. She was not suspicious, and would not believe what was told to her. Her own body and soul in perfect health — without an evil inclination that could be perceived —

how could she believe it — the least part of it? Confidence inspired affection — devotion. The joys of wedlock were dreamed of in a way, for the time being, to transform his nature. The Satanic was forgotten in the glories awakened by the passion of passions. When an old lover made his appearance! New eyes were given him. Dazed at first, he soon saw falsely. Jealousy took possession of him. A scene ensued. He was understood, and dreaded, of course, and there was a separation. The misery that sweet woman escaped!

An old lover appeared.

The third gentleman said he felt some reluctance about telling his story, as it might appear to bear a little hard upon the other sex. But the case was exceptional, and he would be excused. He had met the lady at two dinner-parties, but never at home. He had been struck by her gracefulness and ease of manner, and by her brilliancy in conversation. She had charmed him as he had never been charmed before. He determined to visit her, as they say, with a view to matrimony. The reception was cordial, and he was delighted with the prospect. The beautiful girl was more attractive than ever. Her graceful

The third gentleman's story.

Reception cordial.

person was exquisitely adorned. Her eyes were brighter than diamonds. Tact and intelligence marked her conduct and speech. Her music was finished and chaste: one of her songs touched him particularly: emotion was in every note of it: it reminded him of much that had been delightful in his varied life. The drawing-rooms were adorned in an elegant manner. Mirrors, the costliest, were on the walls. Carpets of velvet softened and warmed the floors. The rugs were pictures. In the midst of his enjoyment it began to storm, and it continued to storm, violently, without intermission. It was a wild night. Far away from his lodgings, he was obliged to accept further hospitalities. The chamber he occupied was in such contrast with the salon he had just left that he was dumfounded. He rubbed his eyes and collected his scattered wits; but he felt lost in the changed situation and conditions. Everything in the room had a neglected look. The draperies were faded and mean. Inhospitableness was in every detail. The bed was most uninviting. The linen was not clean nor fresh. The contents of the pillows were not eider-down by far; and they were lumpy, and had an unwholesome

Tact and intelligence in conduct and speech.

Lost in the changed situation and conditions.

The Truth Revealed to Him

smell. The storm, and the revelation of neglect, and the miserable disappointment, made a very uneasy night for him. The breakfast-room had the same neglected look and the same noisome smell. The carpet had one great offensive spot upon it that had never been forgotten. The muffins and omelet were overdone, and the coffee was muddy. The drawing-rooms, after the night's and morning's experience, appeared affectedly fine indeed, and confused all his memories and previous impressions. He took his leave a wiser but not a happier man. He was sorry to have had the truth revealed to him in such an unexpected way. The thoughtlessness of the imposture had surprised him beyond measure. To call such a household a home seemed a monstrous misuse of the word. Could it be possible for one bred in such an atmosphere to comprehend what a home should be? All idea of cleanliness and comfort had been lost in affectations, disguises, and self-delusion. He frequently met the young woman afterwards, but never otherwise than as an acquaintance: the disillusion had divested her of all attractiveness. The world took possession of him — its cares and responsibilities. Burdens of

An uneasy night.

The disillusion.

others came upon him, one after another, and he believed he was contributing to the common stock of happiness. It was not likely that he would entertain thoughts of matrimony again.

The first lady to speak.

The first lady to speak was strikingly attractive, from her beauty of health and perfection of maturity. She might have stood for Juno in sculpture. She said she would be frank as the rest, and tell her story without let or disguises. The governing cause of her single-blessedness, she said, was discovered by the professor in the examination of her head, when he pronounced her exceptionally small in philoprogenitiveness. Where a bump ought to be, was found a perceptible cavity. When this organ was small, science taught, there would be shown lukewarm attachment for children; they would not be esteemed a blessing; weariness and impatience would be felt in their company; their prattle would not be tolerated. Her experience

She could not abide children.

was in confirmation of science : she could not abide children, except in very rare cases. As studies merely, as a rule, they had been interesting to her. Young animals of other species were about as engaging. As she could not help this perversion, she had yielded to it reluctantly. It was not pleasant to be out of nature in such an extraordinary way. It made marriage — the haven of happiness to most women — impracticable to her. She had dreamed, time and again, of maternity — of being surrounded by her own children; and the joy of relief upon awaking was spasmodic. Other loves than those of motherhood had been vouchsafed to her. She had two husbands, so to speak, — literature and art. Never a day was long to her with a good book for company. Belles-lettres, in all that it included, was ever fresh and abounding in interest. Life in literature was the life she most relished. She could enter into it or quit it at will. The creatures and personages of books did not need to be petted and flattered: unceremonious usage did not offend them. Pictures she enjoyed, and sometimes painted, in a poor way. Her sense of vision was helped by the pastime. She saw more, the more she drew and colored. The possibilities of tints were a perpetual surprise to her. Sometimes she essayed portraiture, but only in attempts to portray manhood in rare specimens. All of her powers were in best employment at such times. Lines

She had dreamed of maternity.

The possibilities of tints.

of thought in a thoughtful face it was her chief pleasure and ambition to trace. Complexions of women and children were too delicate for her brush, as were all expressions of effeminacy and softness. The bold, the strong, the manly, excited her to utmost effort. So married to print and canvas, what more could she desire? She had had lovers — not a few. One poor fellow adored her, and threatened self-destruction if she did not marry him. Another was diverted in his homage by the fascinations of the card-table. Approaches of others were discouraged as waste of the emotions. Nature had appointed her to a single life. Her destiny had been predetermined from the foundation. The daughters of Erebus and Night were executing the decrees of Nature with inexorable decision. Their ministers, the Furies, had not been necessary; there was no resistance.

Married to print and canvas.

The second lady said that a few facts, simply stated, would satisfactorily account for her voluntary maidenhood. She was the eldest of five children. When she was only ten years old her mother became a hopeless invalid, and the cares of a full-grown woman were suddenly imposed upon

The second lady.

Self-Sacrifice

her. She gave up all — head, heart, hands — to her mother and brothers and sisters. The youngest was a mere baby, and you must know the constant attention he exacted. Her father was kind, perhaps, in his way; but he was a confirmed hypo- chondriac, forever groaning and complain- ing of everything. God and nature were at enmity with him, he said. Smileless and discouraging, his presence was a perpetual blight. He never said a generous, inspiring thing to any one of them that she remembered. Unconsciously selfish, his whole thought was of himself and his imagined distresses. The looking-glass was his great resource in his absorbingly self-pitying moods. He would pull at his beard and penetrate the lines in his face, and sighingly wonder what other tortures were in reserve for him. Any misfortune or crisis in the family, instead of stimulating his humanity and sympathy, only increased his malady. When his wife suffered most, he was most jealous of attentions to her. He bemoaned himself and groaned, when a little bit of self-sacrifice and tenderness would have brought sunshine into the joyless household, and lightened all its burdens. The baby died when

Her father a hypochondriac.

Bemoaned himself and groaned.

he was scarcely three years old. The blue-eyed cherub! His death was a great blow. Her cares were lessened by it; but there was an aching void. A record of the solemn entombment was in everything about her. Special remembrance of him always occasioned a pang. Strange to say, the death of the little fellow seemed to give relief to his mother, and she grew perceptibly better, though still bed-ridden, to remain so till she died. He is better off, she would quietly say, with a touching smile of self-consolation. The girl-children were lovely, and grew in helpfulness. There was nothing they would not do. The boy was always manly, and rapidly developed the most genuine traits. He seemed preternaturally strong and wise. His hopefulness and sturdy self-confidence gave joy to them all. He acquired and thought, and every day grew in intellectual stature. You shall see what will be done for you, he sometimes proudly and heroically said. The world soon recognized his abilities and manhood. His advancement was steady and sure, and he soon ranked an exceptionally prosperous man. The desire of his great heart was realized, and the family at home enjoyed more and more

The solemn entombment.

Pride and heroism.

his fostering care. The girls married gentlemen, well-to-do and generous. Their father was indulged and their mother cherished and petted. Ah! the smile of rejoicing that illuminated her invalid face, after all her trials and miseries. A word or two more would complete all that was necessary to relate of her story. Her noble brother and grateful sisters had settled a generous annuity upon her; and her life was as free as that of any woman could be. She was getting the most out of it that was possible to her, and she believed she had no complaint to make of fortune or condition.

Smile of rejoicing.

The third and last spoke with a little more spirit. The preceding statement made her own less difficult. While her experiences had been alike bitter, they had been more tragical. She also had been a victim of circumstances; the miseries of unfortunate marriage had been indelibly impressed upon her. They had been brought home to her in a way to make her hesitate about accepting an attractive offer, in all respects promising. The marriage of her father and mother had been one of blind passion or affection. Friends had urged a postponement, to give a little time

The third and last.

A marriage of blind passion.

for consideration; but both were infatuated, and would not live apart, even for a short season. Her father was handsome and gay; devoted to the world and its pleasures; governed without limit by his impulses. His appetite for drink increased; and indulgence soon became dissipation. Evil associations made him rude and reckless. He changed from what they called a gentleman to a brute. He abused his wife in outrageous ways. The narrator called attention to the mark on her mouth, the same exactly as the scar left on the lip of her mother by the heel of her husband, months before she was born. Daughter and mother with the same ineffaceable memorial of brutality! Her father, she said, had tried to be kind to her sometimes while she was a child; but long before she became a woman everything like affection had disappeared from his conduct. He even hated her, as he did her mother. A complete transformation had taken place. He had grown to be a monster. He seemed to have three faces, like Cerberus, every one of them cruel; and each one had the remorseless evil eye. To get behind him, and to escape the fatal look, was impossible. He

Appetite for drink increased.

Three faces, like Cerberus.

saw all, and suspected more. Physician,
clergyman, friends, male and female, were
objects of his suspicion and jealousy.
You talk about moral atmospheres! Think
of living in one of profanity and drunken- *Profanity*.
ness! Recollections of what she and her
mother endured, terrified her. An in-
cubus was upon their lives, asleep and
awake. Certain demoniac noises and
oaths came to them in ways to threaten
reason. Pandemonium could not produce
worse. From bad to distressing the
wretched days continued; till one night
the monster was brought home dead, with
a bullet in his brain by his own hand. His
poor, relieved, heart-broken wife survived
him a few weeks. Her life went out in
agony. The event of her own marriage,
often talked over with her mother, and
postponed at her request, would be con-
summated in the early autumn. Her lover
was acquainted with every circumstance of
her life, even to the birth-mark on her lip, *The birth-*
and had many times befriended her and *mark.*
her mother at the risk of his existence.
He was a noble fellow, and she dared hope
for happiness the remnant of her days.

Something like a breeze, by this time,
was seen to ruffle the surface of the sea,

a mile away, or less. One said it was a school of mackerel on the way to Arcturus. Howbeit, they made sail; and Zephyrus came gently to fill it, and bear them away to their several hostelries.

DRUNKEN-NESS. Even the Sub-Cœlumites found drunkenness the most stubborn of all the social evils. Though rare, they found it impossible to abolish it utterly. Destroying the effects of alcohol was like annihilating the archenemy. They believed implicitly with the poet, that the loved and hated thing was introduced by Satan into the tree of knowledge before the primal pair partook of it, and was attended with the same effects that had followed it ever since. Confirmed drunkenness they regarded as one of the most virulent of moral and physical diseases, and they took every pains to protect society against it. Some idea may be had of their success by remembering the early excesses of one of the countries that had supplied them with much of their pop-
A matter of history. ulation. They had history for it that on the signboards of noted gin-shops in that country it was announced that a customer might get drunk for a penny, and dead drunk for two-pence, and have straw for

nothing. Faith was kept by providing cellars strewn with straw, on which the customer who had got his two-pennyworth was deposited till he was ready to recommence. Higher, socially, excesses were as extreme, but different. They had the statement of a noble writer that he was present at an entertainment where a celebrated lady of pleasure was one of the party, and her shoe was pulled off by a young man, who filled it with champagne and drank it off to her health. In this delicious draught he was immediately pledged by the rest, and then, to carry the compliment still further, he ordered the shoe itself to be dressed and served up for supper. The cook set himself to work upon it; he pulled the upper part of it, which was of damask, into fine shreds, and tossed it up in a ragout; minced the sole, cut the wooden heel into very thin slices, fried them in butter, and placed them round the dish for garnish. The company testified their affection for the lady by eating very heartily of the impromptu. The authorities of Sub-Cœlum were prompt to grant a divorcement of man and wife when either became a victim of drunkenness. Hospitals were established for confining and

Statement of a noble writer.

Authorities prompt to act.

treating it, not without hopeful consequences. Licenses to marry were not granted without inquiry as to the habits of applicants and their progenitors. Tendency to intoxication, even, was alarming, and might entail itself. Prevention was the only sure remedy. Indeed, no short list of questions must be answered satisfactorily, under oath, before a license could be obtained. Drunkenness was not the only evil that society did its utmost to cure, to limit, and to prevent. Diseases that rot the moral and physical structure were searchingly hunted out and pursued while a visible remnant of them remained to taint the generations. Habitual lying, hypocrisy, and dishonesty were recognized moral diseases. A deliberate breach of trust was such a monstrous crime in their moral code that the name and blood of the perpetrator were not perpetuated. Society held itself not guiltless if it permitted the odium of serious crime to descend upon the irresponsible and innocent, to say nothing of possible continuance.

Licenses to marry cautiously granted.

Divorce. While there were other legal causes of divorce than drunkenness, the authorities were slow in acting upon them. Separations were oftener authorized than divorces. The theory and rule of their civilization were, that husband and wife must live together, and not be long separated. Estrangement was provided against in every possible way. Trifling differences between married people were not patiently considered. A custom of the olden time became a rule of action in their courts. When a quarrelsome couple applied for a divorce, the magistrate did not listen to them. Before deciding upon the case, he locked them up for three days, in the same room, with one bed, one table, one plate, and one tumbler. Their food was passed in to them by attendants who neither saw nor spoke to them. When they came out, at the end of three days, neither of them wanted to be divorced.

Estrangement provided against.

Victims of occasional intoxication were kindly provided for by the establishment of Refuges, for their care and protection. Fortunately, they were not many, and the wonder was they were so few, considering the exigencies and extremities of even exceptional human life, and that the vast majority there, as elsewhere, were governed by their passions and emotions, and

REFUGES FOR CERTAIN OCCASIONAL VICTIMS.

not by their judgment. Reason, there, as everywhere, was the property of the chosen few. Living to-day upon the experience of yesterday, and so providing for the morrow, if it come, was easier of philosophic statement than practice. The faculty of continuing in the right way, without being once in a while turned aside by folly or temptation, was not given to common mortality: it was a rare endowment — the gift of God. Their stream of life, also, had its numberless eddies, to obstruct and hinder. Caught by them, and whirled about, it was difficult to get themselves back into the current the same creatures as before, to enjoy again, in the same healthful way, the inspiration of progress. Maelstroms, indeed, they sometimes proved to be, wrecking hopelessly, if not utterly swallowing up, the moral man, in their unconditional irresistibleness. Human wisdom, as they possessed it, was largely the result of suffering and blundering. It was not given to them to know the next step but by taking it. Discouragements and calamities made them timid about taking it at all. Business complicated and embarrassed, they could not always see their way to solvency. Expenses exceeding income, ruin

The faculty of continuing in the right way.

Suffering and blundering.

impended. Fraud victimized and paralyzed them. Conspiracy gave them new eyes. Immoralities were in danger of being exposed. Losses, one after another, seemed ruinous altogether. A spendthrift boy brought unexpected entailments. A foolish girl wounded the family pride and compromised her honor. Domestic infelicity was possibly creating new irritations. A rasping voice and intrusive nose might never be out of his ear and affairs. Superadded, a dismal atmosphere, to overwhelm with gloom. What more natural, even in Sub-Cœlum, than a short cut to temporary relief through the bottle? A little of the artificial sunshine being found good, a flood of it was better, and intoxication ensued. Days of it, probably, before discontinuance was thought of. The poor victim — perhaps for the first time in his life — cares not to go home: he goes of preference to the Refuge, where he is admitted upon application; few questions are asked; discipline is so slight as hardly to be felt; he is thoughtfully let alone; permitted at will to wander through the beautiful grounds, without molestation; supplied with everything necessary to his comfort, in the way of food, baths, and

A spendthrift boy.

Few questions asked.

clean beds; but not a drop of anything intoxicating is given to him during his stay. The healing solitude and absolute freedom, in a few days, complete his restoration. No record is made of the matter, and he is discharged without scrutiny or pledge. So little indeed is made of the circumstance that Gossip herself is hermetically dumb concerning it.

The healing solitude.

Retreats for convalescents were established, here and there, throughout the Commonwealth. People came to them from every part, — especially those who had not comfortable homes. These Retreats were situated in attractive places, where the air was the best, and where inviting accessories could be easily provided. Trees were planted of the most beautiful varieties. Flowers in abundance were cultivated. Fountains played, in volume and spray, displaying rainbow colors to the greatest advantage. Rills ran through the grounds in a natural manner. Ingenious little contrivances for entertainment were operated by them. Mechanical skill exerted itself to invent diminutive engines for all sorts of purposes. Musical instruments were made to play by the force of

Retreats for Convalescents.

Fountains played in volume and spray.

A Convolution of Rainbows 141

the element. The prettiest little ponds were provided for the fishes, and for the birds to bathe in. Of the former, those of every brilliant color were to be seen; and of the latter, those of every quality and tint of plumage. In moulting time the birds were especially interesting. When the sun shone, the atmosphere, at times, was a convolution of rainbows. Intelligent monkeys climbed about in the trees, and suspended themselves by their tails. Grave and gay, wise and foolish, they never ceased to be objects of study. Record was made of their cunning and imitativeness, and they were respected in proportion as they were known. Lessons were taught by the application of their powers. Not every man was exalted in comparison with them. Their ailments — much the same as those of their human brethren — were treated not empirically, but scientifically — too much affection for them being felt to permit mere practice upon them; besides, they might avenge themselves, — curious instances of the kind being of record in all the institutions. Milk of the cow and the goat and the mare was supplied as needed. The cooking was exactly adapted to the stomachs and nerves and

The fishes and the birds.

Life in the trees.

Milk of the cow and the goat and the mare.

palates of the feeble. The most delicate dishes were served to nourish and stimulate. Sleep-producing qualities were specially aimed at, — the belief being prevalent that frequent and complete suspension of the functions of the hemispheres of the cerebrum was necessary to sound physical, intellectual, and moral health. Drugs were eschewed, as especially for the hospital. Generous wine, in sufficient quantities, was supplied, but nothing stronger. Tea, also, and coffee, were forbidden, except under peculiar circumstances, the excessive use of either being held accountable for many idiopathic and morbid conditions. Manifold amusements were provided, — such as were suited to the tastes and strength of convalescents. Only visitors were admitted who were healthy; and those must be considerate and of stimulating effluence. The brooding mood and complaining habit were shut out as pestilential influences. Full veins and abounding vigor were welcomed as inspirations. Sickness and death were not subjects of conversation. Restoration to health being the object of these wise and merciful Retreats, anything to hinder or thwart that was scrupulously forbidden. Inmates must get well, and not

Drugs eschewed.

Only healthy visitors admitted.

expend any part of their powers, moral or emotional, in brooding over distresses and perils past and escaped. Reluctance to adopt cheerful moods, and to coöperate with wise and compassionate treatment, were grounds of prompt dismissal from the institution.

Hospices for visiting strangers were in all the considerable towns. They held a place half way between the hotel or hostelry and the private home. They were conducted respectably but not extravagantly. The strictest cleanliness was observed, and plain food was generously furnished. There, as everywhere, pains were taken in the preparation of articles to be eaten; nothing was spoiled in the cooking. Abundance of pure water was supplied for bathing purposes. Accessible reception rooms were provided. The prices charged were only a trifle above the cost of material and service. The social character and habits of the people required such institutions. Enjoying abundance of leisure, a good part of their time was taken up in visiting, and every facility was necessary to free intercommunication. From town to town they went, singly and in parties,

Hospices for Visiting Strangers.

Such institutions required.

and these Hospices were comfortable enough homes for them while they remained. Their friends were relieved of the burden of entertaining them, and never wearied of seeing them. The absolute freedom all enjoyed was favorable to happiness. Housekeepers were relieved of anxiety and a great part of the social pressure. It was astonishing the amount of pleasure received from this free intercourse with visiting friends and strangers. Nobody was embarrassed by obligation. All material enjoyments were paid for. Politenesses were voluntary, and without complications. Society had almost nothing of the debt-paying element in it. Pretenses of overwhelming gratitude and favor were without excuse, and were not exhibited. A thousand and one of the little insincerities and hypocrisies were avoided. Disguises, so many, were not thought to be necessary to appear kind and hospitable. It was possible to look into each other's faces without embarrassing remembrance of deceits and dissimulation. Self-respect was less difficult when free of the burden of petty sins against veracity. Greater transparency existed in the social relation. Less of conduct was a mockery of con-

Absolute freedom.

Disguises not necessary.

science and religion. Young people, especially, were benefited by the freedom and liberal facilities. With the aid of the public Hospices they saw each other often, and in a catholic manner. Life was not so much a game with them. The sexes were upon a common plane. They were more apt to comprehend each other, and be better fitted for the holy bonds. Freedom from much expense and ceremony gave more time and better opportunity for consideration; and precipitation in marriage was not the rule by any means. In the enlarged facilities for intelligent courting, society found important protection. There was less likelihood of crazy infatuation. If the suitor was not the right kind of gentleman, his sweetheart was pretty sure to know it. His conduct was more open to inspection, and would expose itself, if not based upon trustworthiness. In the general interchange, outside of business relations, the Hospice was found indispensable. Greater opportunity was given to the offices of patriotism, charity, and benevolence. Society was more like a great family. By its liberal and healthful intercourse, its civic and social virtues were perpetually nourished.

The sexes.

Like a great family.

146   Sub-Cœlum

INVENTORS AND SCHOLARS. Inventors and scholars, in a pecuniary sense, were not apt to be more prosperous there than elsewhere, and so were relieved of many ordinary burdens. Society, having been benefited by their labors, was willing to compensate them as it could. In cases where they had grown old and poor special provision was made for them. Especially they were preferred for any public service they could perform. Considering the great intellectual activity, the wonder was there were not more that required assistance. The proverbial unthrift characterizing the purely intellectual classes they had their share of, but no more. You heard the same incidents of innocency of the arts of trade that literature has been recording ever since living and language began to improve. How, while they were evolving great
Cheated of their pennies. thoughts, they were cheated of their pennies. The same old instances of forgetfulness of self and material interests that ignorance is forever quoting to fortify its self-conceit. A man had actually died while reading a proof-sheet of great astronomical researches, when not a crumb to eat was found in his lodgings! Defective, half-made creature, of course, not to pro-

vide properly for his stomach! Jones, who had a great estate, did not care for constellations and comets. Smith had accumulated, and hardly knew how to read! What of all the host of stars? Cabbages did not grow better for all the knowledge of them. Incompatibles, they said, were thrift and scholarship and scientific investigation. Intelligence understood the matter better, and provided in many ways for neglects and omissions. When manufacturers made great fortunes by utilizing great inventions, whatever the terms or circumstances of purchase, they did not forget the inventors. If they did not remember them fittingly and substantially, Government prompted them by significant means. They were required to furnish money or employment — assistance to the inventor being as far as possible in just proportion to the pecuniary value of the invention. Publishers, in case of unexpected large sales of publications, were expected and required to further share their profits with authors, if necessitous. The prevailing sense of justice amongst appreciating people did not permit a neglect of classes preëminently worthy. Conscience was wide awake in such cases.

What of all the host of stars?

The prevailing sense of justice.

OLD PEOPLE AND CHILDREN. Very few old people or children were objects of public charity. The humanity of the people and their religion were against it, except in cases of direst extremity. Affection was more than water, and provided for its own. No greater disgrace could fall upon a man than by the neglect of the old or the young of his own blood. Whatever the exigencies, relief generally came from the natural source. Families were not so large but that room might be made for one person more, in extremity. The aged were guarded and comforted by their children or children's children — by their relatives, immediate or remote. Degrees of relationship were not counted when suffering presented. Blood was not denied in any condition of indigence or affliction. It flowed and interfused unconsciously on occasions of calamity. Re-

Religion more than skin-deep. ligion — more than mere words, and more than skin-deep — delighted in self-sacrifice. The helpless were helped as a religious privilege, and the burden was not shunned nor calculated. The Founder of their religion was the poorest of the poor, and the religion He founded was for the poor especially. Hungry and thirsty, He went about doing good, though rejected and despised. He was love and self-sacrifice incarnated. Pitiful, shameless followers, who deserted their own blood, in poverty or wretchedness.

The most beautiful spots in Sub-Cœlum were the burial-places. The celestial visitant, hovering over, must have been charmed by their attractiveness. Nature and Art did their utmost to beautify them. Grounds were chosen for their diversity and irregularity. What Art did was only to assist Nature: not a thing was done to show her tricks and fantasies. Hills and valleys in abundance, little was left to the landscape-gardener but to adorn them naturally. The native forest was little disturbed. Additional trees and shrubs were planted to give greater variety. Exuberant vines crept and climbed about in fantastic ways. Perennial plants and flowers were everywhere in view, and different at every turn. Exotics were cultivated where not too much labor and expense were involved, and where they did not give a look of too great artificialness. Particular pains were taken in the cultivation of plants and shrubs the leaves of which emitted pleasant perfumes; rosemary, lavender, sweet-brier,

Burial-places.

Perennial plants and flowers.

and the like; which, upon the slightest touch or disturbance, filled the air with delicious odors. Roses, roses, were everywhere; and pinks, too, in great abundance. Sinuous roads and walks ran in and about bewilderingly. The line of beauty was conspicuous. The birds delighted to dwell in these enchanting places: they were fed and cherished in every hospitable and affectionate way. Squirrels of many varieties were perfectly domesticated, and added greatly to the general animation. They came down out of the trees to be noticed and petted. The children they delighted to run over, searchingly and caressingly. Names were given to the prettiest, and when they died they were mounted or decently buried. These lovely burial-places were freely visited by everybody without distinction. No embarrassing rules or by-laws were placarded on the gate-posts or elsewhere. No scrutinizing look was given by officer or lodge-keeper at the entrance. It was only expected that the sacred place be not made a haunt, and that good behavior would characterize the conduct of the visitor, such as enlightenment and good feeling would suggest as befitting. The public was encouraged to go to

The line of beauty conspicuous.

No embarrassing rules.

Death Leveled All

the beautiful cemeteries for their civilizing, refining, and moral influence. If an adult or half-grown person misbehaved in one of the resting-places of their dead, he was uncivilized, and hardly responsible. Breaches of good conduct were so rare as to be historical. Their religion taught immortality, and that death was but emancipation. Believing that they began to be here what they were to be hence — that they made their future in this world and took it with them to the next, they felt the responsibility of living; and anything that tended to increase that feeling was religiously encouraged.

Breaches of good conduct rare.

There was little distinction exhibited in marking the graves. The stone-cutter was not required to be an artist. Costly tombs and monuments were not in fashion. The graveyard was considered a poor place to draw the lines upon penury. Wealth was too considerate to display itself in places of the dead. Fortunes were not expended in commemorative columns and shafts. Ambitious display stopped short of the tomb. The poorest man was not reminded there of his indigence — grandeur did not mock him at the grave. Death leveled all.

LITTLE DISTINCTION IN MARKING GRAVES.

Sensibility might show itself, but not cash. Adornments were such as affection suggested to thoughtfulness and refinement. Any one might embellish a grave. Plants, flowers, a modest stone, intelligent care, were not costly. Simplicity and tenderness gave greatest distinction. Birds were sometimes lured, by ingenious and affectionate means, to nest on the graves. Broods of the same pair successively took wing — emblems of immortality. Flowers grew better by the tears dropped upon them, and the fragrance they exhaled was super-terrestrial. Little evidences of affection and remembrance were everywhere to be seen. Lettered phrases were touching to read. Memorial verses from time to time were found in the grass. In every imaginable way the deep humanity and profound religious sentiment expressed themselves in these sacred places. At the same time, nothing gloomy or dreadful was suggested. Remembrances of the dead were quite as apt to be joyful as distressing. Pleasant things of them were rehearsed, and they lived again, and were reënjoyed. Children felt themselves nearer their lost parents by cheerfully reviewing their kindnesses and self-sacrifice.

Emblems of immortality.

Nothing gloomy or dreadful.

Parents forgot themselves in agreeable reminiscences of their children. The best was remembered as most apt to be perpetuated; the regretted was buried with the dust. The common distinctions were forgotten in these cities of the dead. Costly improvements in roads and chapels were directed by the general management. Opulence was arm-in-arm with indigence in the enjoyment of the pervading beauty, and nothing existed to suggest any disparity. In the park or in the public gardens grandeur might display itself, but not among the graves, where all humanity was common dust. Not that pride aped humility: it was humiliated indeed by the thoughtful consciousness of the common lot. *Common distinctions forgotten.*

Neither were ostentatious funerals in vogue in Sub-Cœlum. They did not comport with the prevailing ideas of propriety. It had been a great many years since anything of the kind had occurred there. Simplicity, rather than display, characterized the burial of the dead. Any appearance of vanity or vain show, in connection with death, had come to be regarded as more or less barbaric. Costly equipage and con- *Funerals.*

spicuously fine dress had long since been tabooed. In one of the countries from which a great part of their population was descended, ostentatious and expensive funerals had been the rule. They had record of one, where the procession was a mile long, and walked sixteen miles to the place of burial. Every variety of refreshment was served, and over five hundred gallons of whiskey were consumed. Simple religious services at the house of the deceased were customary, attended by the family and their friends, and such acquaintances as were invited. The remains then passed into the hands of the director, and were quietly conveyed to the cemetery, accompanied by a few near friends. The face of the poor dead human body was not exposed to the multitude at any time. Curiosity was not gratified in so indelicate and rude a way. Mourning, in the sense of outside manifestation, was rarely exhibited. Their cheerful views of life, here and hereafter, led them to accept the inevitable resignedly and hopefully. They could not account for this existence without a belief in a better to succeed it. Fidelity and purity and humanity in this, would be followed by felicity in that. It

Ostentation and expense not the rule.

Their cheerful views of life.

Display Avoided

was a faith they all had, without qualification. Expense they avoided as far as practicable. They regarded the occasion of death as not a fit one for the display of this world's possessions. Good in their way, they were not to be compared with the priceless abundance promised to the worthy. Besides, their delicate sense did not permit them to exceed the average in expensiveness at the last hour. Penury was not to be reminded of its limitations by prodigality. It was a common thing for neighbors to bear each other to the burial-place, and to dig each others' graves.

This world's possessions.

In every considerable burial-place there was a convenient and commodious chapel, adapted to religious and other exercises connected with the dead. There the sealed caskets containing the remains of persons well known were frequently placed, and for a time opportunity was given to the public for free expression upon the lives and services of the deceased. The general intelligence and readiness of speech, with the prevailing habit of reflection, made these occasions particularly interesting and impressive. The utmost propriety and solemnity were observed. Sometimes elabo-

CHAPELS IN BURIAL-PLACES.

rate orations were delivered; but generally remarks were spontaneous and unpremeditated and brief. Incidents of the life that was ended, illustrating its character, were related in a natural, conversational way. Foibles were forgotten in the generous consideration of aims. So much that was good was found to be said, that disparagement had no voice. It was not remembered when an uncharitable thing had been uttered on any one of these occasions. The people were too wise to expect perfection in any human life, and too considerate, if not too good, to cherish memories of common errors and occasional lapses from strictest rectitude. Analyses of character, while often acute, were always kindly and forgiving. It was surprising how the strong light of observation brought out the virtues. A man, thought by the casual beholder to be hard and ungenerous, appeared, in the judgment and knowledge of his friends, a just and self-sacrificing citizen. So far from being selfish, there was nothing he would not have done for others, without advertising it. Ungraciousness of manner was his misfortune. A poor man, the victim of his appetites, appeared a noble fellow in instances where he had risked

Foibles forgotten.

Ungraciousness of manner.

Essential Unselfishness

his life for the helpless. His depravity had exhausted itself upon himself. Tributes to his humanity and essential unselfishness were in hearts without tongues to express them. An eminently proud man to superficial apprehension, in the flood of truth poured upon him, seemed only the self-respecting gentleman. His pride indeed was lost in his profound integrity. An unfortunate woman, a martyr to her beauty, who had incurred the odium of her sex by certain irregularities, lay one day in this house of the dead, as might be thoughtlessly supposed, for condemnation. Far from it. The silence of the tomb was broken by feminine sobs, and the best of her sex repeated, He that is without sin among you, let him first cast a stone at her. The whole house rose responsively, and passed out, one by one, touched by the spirit of the Master. Men and women, conspicuous by their acts of patriotism and humanity, received their just tribute. But nothing fulsome escaped the lips of any one. It was considered a great offense to say of the dead what could not be truthfully said of any living human being. Acts were recognized and appreciated; but motives were not discussed.

Only the self-respecting gentleman.

Nothing fulsome escaped the lips of any one.

MOTIVES.

Interesting and surprising.

Both satirized and dignified his conduct.

In that land of intelligence, observation, and introspection, it was profoundly realized that an attempt by law-makers to define motives, and by judges to punish them, would be puzzling occupation. To the self-observant Sub-Cœlumite nothing was more interesting and surprising than his own, as they appeared to himself, and as they were interpreted by others. Often they seemed wholly beyond his comprehension or control. They were prompted he did not always know how nor why, and would lead him he could not tell where. Their meanness often humiliated him, and he used the utmost caution and carefulness to conceal them. His complacency was only preserved by a consciousness of the world's ignorance of them. Better motives than the real ones were often attributed to him, which both satirized and dignified his conduct. His greatest achievements often sprang from motives so insignificant that he would have been ashamed to acknowledge them. His apparent and exemplary virtues would have lost much of their effect if the secret vices which alarmed them into exercise were exposed. Worse motives were also found for his conduct than ever entered his heart,

The Protecting Statute 159

the possession of which would have made him a different man. If conspicuous good to others resulted from an act meant primarily to benefit himself, his sagacious benevolence was praised and his character accepted a model. If wrong was incidentally or intentionally done his neighbor through his neighbor's simplicity or ignorance, his conscience was soothed by the protecting statute. He had been annoyed by an ostentatious recognition and acknowledgment of acts, with a parade of assumed systematic intentions, when the real ones so spontaneously sprang from his humanity that design or calculation was impossible. Their intrinsic goodness was so disparaged and obscured by misinterpretation and flaunting that their promising fruit was stinted in the growth. The sweeter virtues, crushed into life, are embarrassed by being displayed. The silent tear which attends their birth drops away in shame at being discovered.

His sagacious benevolence praised.

There were professional funeral orators, and writers of obituary notices, whose services were frequently solicited. Facts and incidents supplied them were responsibly employed, simply or elaborately as re-

Funeral Orations and Obituary Notices.

quested. Where the character justified it, acute and thorough analysis was made. When connected in any conspicuous way with the public, acts were arrayed and events portrayed to impress its value and usefulness. A good man appeared better by the recital of enterprises of which he was an important part. Where his successes fell short of his aims, coöperation was found wanting. His wise and benevolent projects had to wait for favorable conditions and sympathizing coadjutors. The truth of men and women was told without exaggeration or adulation. Whatever of religion was in the life was shown in the portrayal of its enthusiastic humanity and self-sacrifice. What better could be said of it than that it employed and exhausted itself in the service of others? A good woman, who had bred a large family, and led a long life of devotion and self-sacrifice, worn out by care, and weary of her burdens, came at length to what was supposed to be her deathbed. A clergyman thought it to be his duty to call upon her. He asked her if she had made her peace with her Maker; to which she replied that she was not aware that there had been any trouble. Cases like this were used to

Acts arrayed and events portrayed.

A good woman.

illustrate the possible in right directions. Words were slightly estimated in comparison with acts. Canting pretension was silently buried; for what was to be said of emptiness and tongue charity merely? Lives were better than professions. In funeral orations and obituary notices were kindly presented realities; ideals were inferred or suggested. Embodiments of practical virtue and religion stood forth. Standards of conduct were animated by personal illustration, more impressive than didactic instruction.

Words in comparison with acts.

As before said, every one had his vocation and avocation, into which he carried his enthusiasms. By the former he made his money; in the latter he gratified his tastes. Special occupations were numerous, and hobbies also. Favorite objects of pursuit gave full employment to particular faculties. It was expected, in the prevailing mental activity, and dishonor of idleness, that every intelligent person would have some appropriate diversion, befitting his abilities and imagination. Men and women were made more interesting by these worthy pastimes, and were rarely humdrum or commonplace. Their minds,

VOCATION AND AVOCATION.

so to speak, had their little holy of holies, with windows toward heaven, into which they entered in best moods, and recreated their powers. Dull they could not be, stirred so often into definite, ennobling action. It might be only an insect the enthusiast gave his hours of leisure to; but it was an object of creation, and stimulated him. Observation was discovery, and led him into ever-widening fields, and away from the beaten track. Absorbed by his hobby, he was respectful and hospitable to that of his neighbor. He did not apply the epithet hobbyhorsical to any special enthusiasm. He was not found among the dogmatists or satirists. He realized the limits to knowledge, and honored every effort to transcend them. He had been mistaken, and would be again and again. He had laughed, but oftenest through ignorance. If wise enough to understand, he had been awed. Realizing that men are most apt to believe what they least comprehend, he did not require the last fact to give credence. He could disbelieve upon ultimate testimony. Inconsistency or apostasy did not affright him. Modest in his beliefs and disbeliefs, bigotry was impossible to him.

Dull they could not be.

Awed by understanding.

Jupiter and Juno's Wedding   163

These patient and enthusiastic students in particular lines had many opportunities to contribute of their knowledge to the public. They were encouraged to give frequent lectures and demonstrations, which were always numerously attended, and attentively and sympathizingly received. Indeed, these learned talks and exhibitions constituted their highest amusements. They were illustrated in every attractive and ingenious way, and were comprehensible even to the children. The public were proud of these special investigators, who worked for love, and for the general good, and were more than glad to sit reverently at their feet and learn of them. The commonest subjects and objects in the hands of these enlightened enthusiasts became more interesting than any fiction. Crawling and flying things, despised by the common, appeared indeed wonderful in the flash of light ingeniously poured upon them. They say that when Jupiter and Juno's wedding was solemnized of old, the gods were all invited to the feast, and many noble men besides. Amongst the rest came Chrysalus, an Oriental prince, bravely attended, rich in golden attires, in gay robes, with a majestical presence, but otherwise a very in-

Students in particular lines.

Learned talks and exhibitions.

The gods invited.

ferior creature. The gods, seeing him in such pomp and state, rose up to give him place; but Jupiter perceiving what he was, a light, fantastic, idle fellow, turned him and his proud followers into butterflies; and so they continue still, mythology declares, roving about in pied coats, and are called chrysalides by the wiser sort of men. These winged worms, in the hands of a master who had intelligently and zealously studied them, were made more interesting than any Eastern prince in all his splendor of attire and pomp of retinue. Of the more than seven hundred species in one small province, he presented you the most beautiful varieties, in all their gorgeousness and bewilderment of color. Enlarged by the camera, they appeared of enormous proportions — more gigantic than the fabled roc. The wings of certain species, covered on both sides with imbricated scales or feathers, to the unassisted eye presenting the appearance of dust or powder, under the microscope they displayed an arrangement as uniform and characteristic of species as that of the scales of fishes and the feathers of birds. He told you that in a piece of mosaic work there might be nine hundred separate

Turned into butterflies.

Wings of certain species.

Metamorphoses

pieces in an inch square, while the same extent of surface in a butterfly's wing contained from one hundred thousand to nine hundred thousand of these wing-scales or feathers. You saw the despised caterpillar in all his metamorphoses, from the process of hatching — eagerly eating, eating, growing prodigiously, changing its skin several times, evacuating his intestines, suspending himself by a little rope of silk to the under surface of a leaf, appearing, after other mysterious changes, the perfect butterfly, sipping honey from the flowers, like his cousin the hummingbird; reproducing himself and dying, like every other thing of mortality. At the theatre and the circus the people were not amused and profited as they were at these astonishing platform exhibitions. They laughed, and were wonderstruck.

The despised caterpillar.

Even the pestilent, friendless rat was the theme of discourse at one of these popular assemblages. The humane investigator had made a particular study of the animal, and surprised his audience with the number and character of his facts and observations — original and from authentic sources. The nature and qualities of the

SUBSTANCE OF A LECTURE.

creature were presented in a manner to excite astonishment and sympathy. At the risk of being considered tedious, some of his facts and anecdotes are repeated.

Incident related by a clergyman. He related an incident communicated by a clergyman, to prove that the detested rodent shows a consideration and care for its elders on the march which was worthy of human philanthropy. Walking out in some meadows one evening, he observed a great number of rats migrating from one place to another. He stood perfectly still, and the whole assemblage passed close to him. His astonishment, however, was great when he saw amongst the number an old blind rat, which held a piece of stick at one end in its mouth, while another had hold of the other end of it and thus conducted its

A kindred circumstance. blind companion. A kindred circumstance was witnessed by a surgeon's mate. Lying awake one evening in his berth, he saw a rat enter, look cautiously round, and retire. He soon returned, leading a second rat, who seemed to be blind, by the ear. A third rat joined them shortly afterwards, and assisted the original conductor in picking up some fragments of biscuit and placing them before their infirm parent, as the blind old patriarch was supposed to be.

A Necessity of his Existence

Incredible as the story might appear of their removing hens' eggs by one fellow lying on his back and grasping tightly his ovoid burden with his fore paws, whilst his comrades drag him away by the tail, he had no reason to disbelieve it, knowing as he did that they would carry eggs from the bottom to the top of a house, lifting them from stair to stair, the first rat pushing them up on its hind and the second lifting them with its fore legs. They would extract the contents from a flask of oil, dipping in their long tails, and repeating the manœuvre until they had consumed every drop. He had found lumps of sugar in deep drawers, at a distance of thirty feet from the place where the petty larceny was committed; and a friend of his saw a rat mount a table on which a drum of figs was placed and straightway tip it over, scattering its contents on the floor beneath, where a score of his expectant brethren sat watching for the windfall. The propensity of the rat to gnaw, he said, should not be attributed altogether to a reckless determination to overcome impediments. The never-ceasing action of his teeth was not a pastime, but a necessity of his existence. It was explained : the rat had formidable

Ingenious methods.

A propensity explained.

weapons in the shape of four small, long, and very sharp teeth, two of which were in the upper and two in the lower jaw. These were formed in the shape of a wedge, and had always a fine, sharp, cutting edge. On examining them carefully, it was found that the inner part was of soft, ivory-like composition, which might be easily worn away, whereas the outside was composed of a glass-like enamel, which was excessively hard. The upper teeth worked exactly into the under, so that the centres of the opposed teeth met exactly in the act of gnawing; the soft part was thus being perpetually worn away, while the hard part kept a sharp, chisel-like edge; at the same time the teeth grew from the bottom, so that as they wore away a fresh supply was ready. In consequence of this peculiar arrangement, if one of the teeth be removed, either by accident or on purpose, the opposed tooth would continue to grow, and, as there would be nothing to grind it away, it would project from the mouth and turn upon itself; or, if it were an under tooth, it would even run into the skull above. There was a preparation in one of the museums which perfectly illustrated the fact. It was an incisor tooth of a rat,

In the shape of a wedge.

Effect of removal.

which, from the cause mentioned, had increased its growth to such a degree, that it had formed a complete circle and a segment of another; the diameter was about large enough to admit a good-sized thumb. He once saw a newly killed rat to whom this misfortune had occurred. The tooth, which was an upper one, had in this case also formed a complete circle, and the point, in winding round, had passed through the lip of the animal. Thus the ceaseless working of the rat's incisors against some hard substance was necessary to keep them down, and if he did not gnaw for his subsistence he would be compelled to gnaw to prevent his jaw being gradually locked by their rapid development. He quoted from a traveler, whose dogs set upon a rat, and making them relinquish it, he took it up by the tail, the dogs leaping after it the whole time. He carried it into his dining-room to examine it by the light of the lamp, during the whole of which period it remained as if it were dead, — limbs hanging, and not a muscle moving. After five minutes he threw it among the dogs, who were still in a great state of excitement, and, to the astonishment of all present, it suddenly jumped upon its legs,

Curious growth

Feigned death.

and ran away so fast that it baffled all its pursuers. The sagacity of the animal in eluding danger was not less than his craftiness in dealing with it when it came. A gentleman who fed his own pointers observed, through a hole in the door, a number of rats eating from the trough with his dogs, who did not attempt to molest them. Resolving to shoot the intruders, he next day put the food, but kept out the dogs. Not a rat came to taste. He saw them peering from their holes, but they were too well versed in human nature to venture forth without the protection of their canine guard. After half an hour the pointers were let in, when the rats forthwith joined their hosts, and dined with them as usual. Even with his great natural enemy and superior, the ferret, he would sometimes get the advantage by his steady bravery and the superiority of his tactics. A rat and a ferret were turned loose in a room without furniture, in which there was but one window. Immediately upon being liberated the rat ran round the room as if searching for an exit. Not finding any means of escape, he uttered a piercing shriek, and with the most prompt decision took up his station directly under the light, thus gain-

Craftiness in dealing with danger.

Enemies in a room together.

Advantage of the Sun 171

ing over his adversary — to use the language of the duelists — the advantage of the sun. This advantage he managed to keep all through the conflict; when the gentleman, to prove whether the choice of this position depended upon accident, dis- *Not the result of accident.* lodged the rat and took his own station under the window; but the moment the ferret attempted to make his approach, the rat, evidently aware of the advantage he had lost, endeavored to creep between the gentleman's legs, thus losing his natural fear of man under the danger which awaited him from his more deadly foe. A number of rats had got into a basket of grapes, and devoured a considerable part of the contents. The man who discovered them replaced the basket, in hopes that they would again visit it and be caught; but the wary animals never *Wariness.* again came to the basket in which they had been detected. They were so numerous and so bold that they used to come and pick up the crumbs from between the men's feet as they sat at meals. Wishing for a shot at some of them, one of the men dropped a few grains of corn on the ground, and took up his position, gun in hand. Soon one rat bounded across the space as if in great alarm; but no rat touched a grain of the corn, which was exposed for several days and nights, being at last crushed and lost by the passing of feet and vehicles. Rats were numerous in the pig-sties, and ate with the pigs, one of which was turned out of her sty, and a trap-door was contrived to close the trough by pulling a cord. The trough was baited with good maize, of which they were very fond; but neither by day nor by night would a rat venture there as long as the pig was excluded. . Returning the pig to the sty, the rats also returned. A similar case was related in which the rats were so many and so bold that they forced themselves into the troughs, would not be driven away, and consumed no small amount of the food which ought to have gone to the pigs. The owner of the pigs then laid a gun so as to rake the trough, turned out the pigs, and had the trough filled as usual. Not a rat would make its appearance; and at last the pigs were put back, when the rats came trooping in as numerous and as bold as ever. In a gentleman's garden was a conservatory along the roof of which was trained a vine on which the fruit would not ripen; so he had

Exhibitions of caution.

Not one made its appearance.

The Gardener's Discovery

the vine inclosed in a glass frame, in the hope that, the heat being confined, the grapes would ripen better than when exposed to the night air. The plan was successful, and he had a plentiful crop of large-sized bunches. These, however, be- *The big bunches disappeared.* gan to disappear very quickly as soon as ripe, but not bunch by bunch, as would be done by thieves, but only the ripest grapes of each bunch were taken. The gardener, when lying on his back for rest after cutting a lot of branches, heard a scuffling sort of sound, and looking round saw five or six large rats come into the frame; they then jumped up at the lowest hanging branches and managed to knock down two or three grapes, which they proceeded to eat like a squirrel, sitting up on their hind legs and holding the fruit in their front paws. Soon after a large female, followed *A large female with four young ones.* by four young ones, came in, and the old one ran up the vine and bit off one of the ripest bunches, which fell down to the expecting young ones below, who fastened on it and began to eat. The gardener could not keep his laugh, but shouted out, which sent them all away, as if a dog were after them. A lady living in the country had her attention drawn one day to some rats in an outer room, surrounding a pail which had been prepared for the pigs. Observing them carefully, she soon discovered that a young rat had fallen into the pail, and that his friends, to the number of five or six, were in consultation as to the best means of rescuing him. The lady called others of her family to witness their manœuvres, while they continued busily at work, regardless of the presence of the spectators. By twining their feet together — the hind feet of the foremost rat being entwined with the fore feet of the next, and so on — they formed a chain extending over the side of the pail. The foremost rat, supposed to be the mother, then reached down, grasped the young one in her paws, and both were drawn out on the floor. Unfortunately, their deliberations had occupied so much time that the young rat was drowned before he was extricated, and apparently the intelligence of his friends did not extend so far as to attempt resuscitation. Three persons were looking over a garden at sunset, when a rat appeared near a stone wall; then another and another, until five had assembled, the fifth and last dragging a dead rat. A council then seemed to be

Means of rescue.

Resuscitation not attempted.

A Novel Burial

held. Then four of them took the foot of their dead companion and drew the body to a place where the earth was soft. The fifth dug a grave with his head and feet, the depth being sufficient to allow the earth to cover the body. The four afterward assisted in covering it up, leaving the tail of the deceased out of the ground. With a touch of humor, the humane naturalist so far departed from his loved theme as to wind up his interesting hour-and-a-half's talk by referring to the suggestive instance of a mouse and a scorpion being put under a glass together. The mouse was immediately stung by the scorpion, and to all appearances mortally. It remained for some time in a kind of lethargy; but on a sudden it collected its strength, and, as in a fit of frenzy, fell upon the scorpion, killed it, and eat its body entirely up, leaving nothing but the claws. The moment it had swallowed the scorpion the swelling disappeared; no signs of pain remained, and the poor animal was set at liberty, in great health and spirits. Similia similibus curantur.

The fifth dug a grave.

The microscope and the camera were of great service to specialists of every descrip-

MICROSCOPE AND CAMERA.

tion. The revelations of the former exceeded all expectation or calculation. It had been improved until an expert was no longer necessary to adjust it. Intelligent children, even, made free use of it. With their sharp eyes they were constantly making discoveries and noting them. In many houses a room was set apart to microscopy. Specimens without end accumulated in them. Habits of observation were formed, and elevated thinking was encouraged. It was not possible for a man to be groveling and mean whose mind had been trained to atomic observation of nature. No matter what his occupation might be, his diversion made him totally forget it. He was with God in his wonders — lifted out of himself for the time being to a sphere supremely above craft and handicraft. At his bench in the shop the artisan's forehead might be gloomily contracted, and his face appear expressionless; but speak to him of his loved diversion, and his brow lifted, and his countenance was illuminated. Cases were frequently made known where character had been completely changed by the adoption of an elevating hobby. Duality was a recognized principle. Tastes for the sensual and devilish were put aside,

Children made discoveries.

Countenance was illuminated.

The Indian Summer

and finally wholly displaced, by pure and exalting enthusiasms. The camera was hardly less wonderful than the microscope in its discoveries and revelations. Difficulties which for so many years had been insurmountable by the operator and experimenter had been overcome by superior methods. Photographs in colors were common achievements. One color was not more difficult to the camera than another. Effects, indeed, were heightened, as in the Claude Lorraine mirror. Slight color in the cheek became roseate in the picture. Draperies were improved in like proportion. Fabrics appeared finer and richer. Colors were reproduced with super-accuracy. Flowers did not lose much by transfer to sensitized paper. The autumnal forest, in all its colors, contracted to a little space, bloomed and glowed like a great verbena-bed, with the Indian Summer haze enveloping it and the still landscape. Ah! that wonderful Sub-Cœlum season, as one of their great poets described it! The stillness of the landscape in that beautiful time was as if the planet were sleeping, like a top, before it began to rock with the storms of autumn. All natures seemed to find themselves more truly

Difficulties surmounted.

Glowed like a great verbena-bed.

in its light; love grew more tender, religion more spiritual, memory saw farther back into the past, grief revisited its mossy marbles, the poet harvested the ripe thoughts which he would tie in sheaves of verses by his winter fireside.

ELECTRICITY.

Electricity was in general use for all sorts of purposes, especially for transportation and communication. It was applied to every kind of vehicle, and to every kind of machinery. Bicycles, tricycles, and four and six-wheeled carriages, of every description, were moved by it. They ran about noiselessly, as if propelled by the wind. Even the air was traversed by ingeniously contrived vehicles, or balloons. Like the condor, they did not ascend easily or rapidly, but when once up, they sailed away like floating clouds or flying birds, in horizontal curves and straight lines. Above the spires and steeples there was impressive silence; only the song of the lark, and an occasional voice or noise from below, disturbed the profound stillness. People in one talked to those in another. Signals were given by notes of the flute. Telegraphic and telephonic communication was universal. People sat

Above the spires and steeples.

A Great Step 179

in their parlors and listened to oratorios. Lines stretched from farmhouse to farmhouse, and households communicated with ease. Country life was relieved of its loneliness. Neighbors enjoyed each other's music and conversation. In sickness they were advised of every turn. They were guarded against danger. They extended invitations, and gave notice of visits. Offers of help were extended. Horses and implements and vehicles not in use were advertised. Little accommodations and civilities were universal, and closely bound large neighborhoods together.

Country life relieved.

In Sub-Cœlum the people did not snore. They had trained themselves to avoid the disagreeable act. This will not appear strange when it is remembered that in at least one great nation the children do not cry. Harsh, high-sounding respiration was never heard. Their breathing apparatus had been improved by long avoidance of it. Their nostrils had grown like the deer's by habitually inhaling through them. They had learned to keep their mouths shut, except while speaking or eating. Taking the air straight into the lungs was considered unhealthy and unwise. Their

The People did not snore.

ears, too, had increased in sensitiveness by the good habit. The external organ was exclusively relied upon, as nobody opened *Relegated to farce.* his mouth to hear more distinctly. That clownish way was relegated to farce. The women, naturally, had first learned to sleep without making a noise; and, in time, the habit became intolerable in the coarser sex. The former had read of a grand seignior hundreds of years before, and had profited perhaps by the lesson. He kept qualified persons, whose duty it was to travel through the whole empire, to see and choose the fairest and rarest women, having leave to enter all places, nay, their very bed-chambers, to view them in what postures they pleased, but chiefly to know whether they snored or stirred much in their sleep, or whether they slept quietly; and, having made choice, they carried them to the prince, and their parents were much honored and esteemed.

WHISTLING. Nor was whistling heard in Sub-Cœlum. It was a lost art, not worthy of the name. It had departed with other barbarisms, but reluctantly. The tendency seemed to have been born in the people, and was nearly ineradicable. Its stubbornness had been one of the discouraging things in their progress. At first, society laughed at the effort to discourage and prevent it. It was the universal safety valve. As a last resort of impatience and irritation it was regarded as indispensable. Convincing argument was answered by it. It relieved the thinking faculty of vexation. By it audacity announced its defiance, and mean-spirited husbands insulted their patient wives. Nowhere the noise was not heard. It was the gauge of happiness, the standard of ebullient emotion. Nothing showed thoughtlessness like the unpremeditated whistle. The tones of it were the gamut of impulse, and might be marked, as degrees of temperature. Vanity varied them, as it adjusted the drinking-man's hat, hanging it, at last, on his organ of self-esteem. Oft-repeated legends, mixed with religion, had a good influence. It began to be said that the whistler's mouth was not to be purified till after forty days; that the offensive sound was produced by Satan's touching the human body; and that the act was disrespectful to God. Even a whistling noise of any sort scared away the Holy Ghost. A woman tried to coax a dog by whistling, when a pious

The universal safety valve.

Disrespectful to God.

servant interrupted her, Please, ma'am, don't whistle; every time a woman whistles the heart of the Blessèd Virgin bleeds. In some districts it was said that if one *It made the angels weep.* whistled in the evening it made the angels weep. It was a widespread belief — more than a superstition — that it was at all times unlucky for women to make the impious sound, as, while the nails for the Cross were being forged, a woman stood by and whistled. But the thing, perhaps, that had the greatest influence in ridding the nation of the nuisance was a famous instance of heredity everywhere known. The child and grandchild of a persistent whistler were born with mouths puckered, as if in the act; and, as long as they lived, they could only take spoon food, and that by a tube adapted to the purpose. The cases were so peculiar that surgery did not risk attacking them, and they remained a perpetual warning against irreligion and bad manners.

DENTISTRY NOT A PROFITABLE PROFESSION. It was remarkable how generally the people had good teeth. They were lustrous, like ivory, and beautiful to view. It was a rare thing they were lost, except by accident or by wearing away. Dentistry was not a profitable profession. Many causes might be found for this general blessing. The intelligence the people carried into their living was perhaps the chief. Their food, as you have seen, was healthful, and thoroughly cooked. They realized the importance of good digestion, as being *Importance of good digestion.* the basis of all physical, intellectual, and moral soundness. Foods of every kind had been scientifically and practically studied, and their effects accurately determined. Dinners of tragedians, it was said, were adapted to their parts; they ate pork when they had to play tyrants, beef for murderers, boiled mutton for lovers. One of their great poets, seeing another sedulously occupied with an underdone beefsteak, inquired, Are you not afraid of committing murder after such a meal? Much wisdom, they said, was in olives, and that soup and fish explained half of the emotions of life. Fries had been utterly banished from *Fries utterly banished.* the Commonwealth. Thorough mastication was considered a necessity to health, and rapid eating an offense against decency. The pigs, even, had been trained to something like moderation in feeding. The people sat long at table, with abundance of good talk, and kindness, for sauce.

They were ashamed of indigestion, knowing very well that it meant excessive indulgence. Admitting it was advertising their intemperance. Eructation was disgusting and unpardonable. For the teeth, especially, sound digestion was considered better than any dentifrice. Deleterious drugs had not been used for very many years. There had not been a case of salivation in all that time. Devices to sweeten the breath of maidens were not known, for the good reason that they were not needed. The air they exhaled in respiration was as sweet as zephyr in a garden of roses. The breath of kine was not to be compared with it. It was more like a cherub's in perfection of fragrance. Cleanliness was the thing of all things they relied upon. It extended not more to the care of their teeth than to everything pertaining to their living. It was a large part of their religion. Purity was not more shown in their complexions and conduct than in their shining teeth and lustrous great eyes.

Eructation unpardonable.

The reader, following the writer thus far, has inferred the general fondness for squirrels. They were the universal pets of the people. Their liveliness commended them,

FONDNESS FOR SQUIRRELS.

A Suggestive Lesson 185

and their remarkable cleanliness. To see them airing, sunning, and inspecting their beds, bit by bit, was a suggestive lesson in housekeeping. Insects or vermin found no quarter with them. The climate was favorable to the interesting little animal. The native species were many and attractive. Others were acclimated and domesticated without much difficulty. Even the great Malabar squirrel, thirty-three inches long, and as large as a cat, was transplanted successfully. The tendency of the common species to trouble the nests of birds diminished with the care they received. The predatory in their natures was largely eliminated by humanizing influences. Like many other animals, they betrayed a liking for children. While they did not permit themselves to be handled to any great extent — their self-respect prohibiting that — they were very free to run over the persons of those that they liked; peering into pockets and perching themselves on shoulders in familiar ways. Occasionally, in favorable seasons, the squirrels of the forest would multiply so abundantly that days were appointed to hunt them. Only at such times were they generally killed and eaten. In summer they seemed particu-

Climate favorable.

Familiar ways.

larly to delight in the fruit-trees at the sides of the roads. They ran from tree to tree as if in a racing contest with the passing wheels. The alertness of their movements and cheerfulness of their bearing were so inspiring that no wonder the little fellow was a favorite. With his prominent eyes and broad head he seemed to see and comprehend everything about him. The cleft upper lip gave an amused and affectionate expression to his animated face. The soft fur was always clean, and free of any disturbed look; and his long, beautiful tail, expanded laterally, and carried elegantly over his back, was a picture of lightness and grace nobody tired of seeing. There was nothing of the snarling or threatening in his appearance or conduct. He was the embodiment of cleanliness, cheerfulness, gracefulness, and good humor, and was a perpetual inspiration to his biped sympathizers. One of the amusements at the ponds was to set him on a bit of wood and see him floated about by the breezes. With his tail for a sail, he appeared the ideal navigator.

A racing contest.

Great respect was paid to the monkey by the humane inhabitants of Sub-Cœlum.

RESPECT PAID TO THE MONKEY.

The Simian Species

No small proportion of their population had descended from countries where he was an object of worship, and was raised to the rank of a god. Gorgeous temples were erected,

Gorgeous temples.

With pious care a monkey to enshrine.

History describes one of great magnificence; it was fronted by a portico for receiving victims sacrificed to it, which was supported by no less than seven hundred columns. Hospitals were erected for their benefit, where thousands were kept in fancied ease and indulgence. One of the cities, upon its surrender to an invading army, contained a population of forty thousand, and as many monkeys. Specialists in Sub-Cœlum were interested in observing the simian species, and noting their peculiarities. The belief was by no means limited that the human race was descended from the monkey. One species, at least, of ape, was entirely destitute of tail. Students of the animal monkey had collected a great number of interesting facts, showing his resemblance in conduct and traits to the animal man. One female went out to service, made the beds, swept the house, and so far assisted in the cooking as to turn the spit. One on board a man-of-war

Destitute of tail.

assisted the cook and turned the capstan, and furled sails as well as any of the sailors. Monkeys had assisted in tea picking in countries were tea was produced. One pious fellow, like many of the religious castes of his country, entertained an antipathy to an indiscriminate use of animal food, and would eat neither of the flesh of the cow or hog; sometimes he tasted beef, but never eat of it. The young of one species were tended with greatest care, the females having been seen to carry their children to the banks of a stream, wash them, notwithstanding their cries, and wipe and dry them in the most careful manner. A certain specimen would open a chest or drawer by turning the key in the lock, would untie knots, undo the rings of a chain, and search pockets with a delicacy of touch which would not be felt until the thief had been discovered. On board ship an attempt being made to secure an orangoutang by a chain tied to a strong staple, he instantly unfastened it, and ran off with the chain dragging behind; but, finding himself embarrassed by its length, he coiled it once or twice, and threw it over his shoulder. In making his bed he used the greatest pains to remove everything

Assisted in tea picking.

Unfastened his chain.

Human-Like Expression 189

out of his way that might render the surface on which he intended to lie uneven; and having satisfied himself with this part of his arrangement, spread out the sail, and lying down upon it on his back, drew it over his body. Sometimes the captain preoccupied his bed, and teased him by refusing to give it up. On these occasions he would endeavor to pull the sail from under the captain, or to force him from it, and would not rest till he had resigned it; if it was large enough for both, he would quietly lie down by the captain's side. He preferred coffee and tea, but would readily take wine, and exemplified his attachment to spirits by stealing the captain's brandy bottle. He would entice the boys of the ship into play by striking them with his hand as they passed, and bounding from them, but allowing them to overtake him and engage in a mock scuffle, in which he used his hands, feet, and mouth. He never condescended to romp with another monkey on board as he did with the boys of the ship. Persons who aided in killing a red orang-outang, stated that the human-like expression of his countenance, and piteous manner of placing his hands over his wounds, distressed their feelings, and

Must first be satisfied.

Romped only with the boys.

made them question the nature of the act they were committing. A checked shirt was frequently thrown over a specimen, which he wore with great complacency. One day a gentleman wearing linen of a similar pattern appeared in the room, and was immediately singled out, nor was the animal satisfied until he was allowed to examine the shirt, pulling it out from the breast, and holding it in comparison with that which covered himself, expressively looking up in the gentleman's face, as if doubtful of his right to a garb which agreed so nearly with his own. One said of monkeys as a dish that they were excellent eating, and that a soupe aux singes would be found as good as any other, as soon as you had conquered the aversion to the bouilli of their heads, which looked very like those of little children. Very remarkable, they said, and curious beyond measure, were the seeming consciousness of evil and apparent instinct of Satan that these very human animals, under certain circumstances, exhibited. Turtles and serpents were sometimes put into the cells of poor captives. They did not much care for the turtles, but the snakes were the very devil.

A curious instance.

Instinct of Satan.

The dog, next to man, was esteemed for his companionable qualities, and for his integrity. His estimable nature was recognized and appreciated, and was developed in every way that was practicable. Kindness and encouragement did for him what it did for humanity. Treated like a dog was not a saying in that country. Bad dogs were not more numerous than bad men, and were as mercifully treated. Only incorrigibleness cost them their lives. Hopeless depravity in man or dog was guarded or punished as humanity willed or permitted. Cruelty was for savages. High qualities of the animal were as well comprehended as those of man. His fidelity had ever been proverbial. Other animals acknowledged kindness, but were incapable of voluntary sacrifices. Only man and dog spontaneously risked their lives in the service of others. A portion of the population were of a race of affectionate and polite savages, who claimed their descent directly from a dog. They were described by the traveler as having low, musical voices, with a smile full of sweetness and light. So descended, the animal was their close friend and associate. He was taught to do many useful, graceful, and

Qualities and Faculties of the Dog.

Only man and dog self-sacrificing.

generous things; but especially he was used as a guard and protector. In one of the churches on Mount Athos was a fresco representing Saint Christopher with a dog's head. Many instances were related of his fidelity to the point of death. He was pronounced a true philosopher by the greatest of philosophers, because he distinguished the face of a friend and of an enemy only by the criterion of knowing and not knowing. Whenever he saw a stranger he betrayed mistrust; when an acquaintance, he welcomed him, although the one had never done him any harm, nor the other any good. He determined what was friendly and what was unfriendly by the test of knowledge and ignorance. It was a saying that when you go to visit a friend at his house, you can perceive his friendliness the moment you enter the door, for first the servant who opens the door looks pleased, then the dog wags his tail and comes up to you, and the first person you meet hands you a chair, before a word has been said. Intelligence and cordiality were much the same in man and animal. Dog Wheat was not a perfect dog; he had his aversions, as men have. He snatched cats, and they fell dead. But

A true philosopher.

Knowledge and ignorance.

Dog Wheat.

he was magnanimous towards his own; he took the part of small dogs, and of dogs that were muzzled. Two friends, man and dog, went out for a walk together. The latter had contracted a deep cold, and suffered, on the road, two or three violent paroxysms of coughing. Returning to the village, master, or superior, had occasion to go into a shop where sweetmeats and candies of all kinds were kept for sale. While passing a word with the proprietor, something was heard to fall upon the floor a few paces away. Turning round he discovered that Diogenes had reached up and knocked down a package of medicated candy —marrubium vulgare— and was eagerly eating it. He knew what was good for his cough. A faithful but sinful dog, misnamed Pluto, had been betrayed by his immaculate master into the hands of an executioner. When the unhappy creature comprehended his hopeless situation, and just before the fatal axe crushed his perverted brain, he gave his false friend a searching, miserable look, as much as to say, What has Pluto done to you, that you should betray him to death in this perfidious manner? The astonished, appealing expression of discovery and rebuke haunted

Out for a walk.

Diogenes's wisdom.

Pluto's false friend.

the conscience-smitten owner in hours of disturbed sleep and wakefulness. It infixed itself in his memory, it distressed his soul. The incident was made public, and ever afterwards the killing of the canine species was determined by Council.

<small>HORSES BRED FOR MORAL QUALITIES.</small>
Horses, as said before, were bred for moral qualities, rather than for speed and strength. Good temper and trustworthiness were prime considerations. They were treated with great kindness, and were trained to many valuable and ornamental uses. Breaking, or violent usage of the young animal, was not known. His spirit was not crushed, but cultivated, along with other good qualities. He was found to be good as he was well treated. He grew in beauty, also, under affectionate care. Horsemanship was a favorite amusement of the people. The beautiful shaded roads invited and encouraged it. Rivalry was general in all fitting feats and exercises. Ladies and gentlemen were ambitious of distinction in them. Men did not allow themselves, Mazeppa-like, to be bound to wild horses, and let loose on the plains and roads; nor women to represent Godiva, with flowing hair and close-fitting suits.

<small>Rivalry general.</small>

Grace gave distinction rather than daring or boldness. Beauty on horseback was supereminent, and received homage. Poetry in motion was a fair woman and her proud palfrey so perfectly matched that, centaur-like, they appeared and moved as one, unconsciously, semi-human and semi-equine, — tasting in fullness, in the master's language, purest life as it came from the bosom of the deities. Amphitheatres were not uncommon, where displays were made in horsemanship. Horses were trained to perform graceful evolutions, circumpositions, and convolutions, and to enjoy them. The circus was a favorite place of entertainment for the people. What will and the human body could not do, was a never-ending problem of interest. They were proud of their bodies and their minds, and liked to see them tested coöperatively, especially in equestrian exercises. The superior intelligence of their horses was illustrated in the reply of the distinguished rider, when asked if there were not times when, from physical or other causes, he felt doubtful about being able to perform his difficult feats. Yes, he said, there were such times; but his horse always knew of them! Aware of his increased responsi-

Centaur-like.

Reply of a rider.

bility, the noble animal was more than ever thoughtful and circumspect — accommodating himself carefully to his rider, being always exactly at the right place at the right time. Famous horses, grown old, were not neglected as in other countries. The fastest mile horse of his day, in one of them, was consigned to a coach, and at length was found in a ditch, stoned to death. Another, as celebrated, was drawing a cab, after having won seventeen races. The religion of the people, as well as their humanity, forebade such brutality. Happily, they were not insensible to pity or shame.

At the right place at the right time.

LOVE FOR BIRDS.

The birds, of course, were favorites of this enlightened, tasteful, and kindly population. They recognized in them many of the same qualities and traits they possessed themselves, and delighted to study them. Of the more than eight thousand known species they enjoyed a generous proportion. They were not so far away from the equator but that they had many of the most beautiful tropical varieties. The superabundance of the flowers invited them, especially the humming-bird. Over one hundred of the more than four hundred

Wisdom of Birds 197

species of that interesting family, from the smallest to the greatest, were found within their borders. Even the little flame-bearer — sometimes found inside the crater of an extinguished volcano — was occasionally discovered. Its scaled gorget was of such a flaming crimson that, as a naturalist remarked, it seemed to have caught the last spark from the volcano before it was extinguished. It seemed to prefigure the refinement and glory so often resulting from complete self-sacrifice and devotion to the worn-out and helpless. The wisdom of the little birds interested them. Mention was made of a nest of one beautiful species, which, being heavier on one side than on the other, was weighted with a small stone to preserve the equilibrium. They did not permit the wanton destruction of the humming-bird, or other varieties of birds of bright plumage, for mere decorative purposes, as less enlightened peoples had indulged, to the almost entire extinction of many genera. Their experience had taught them that all birds were useful, and they referred to their perfect and abundant fruits and grains of every kind as evidence of it. The great bird of paradise, so rarely found in any other part of the

The little flame-bearer.

Great bird of paradise.

world, was not uncommon in Sub-Cœlum. The splendid ornaments of this species were entirely confined to the male sex, the female being a very plain and ordinary bird; though the young males of the first year so exactly resembled the females that they could only be distinguished by dissection. Whence these philosophical people deduced an argument for limiting coeducation! The fact that the ordinary bird of paradise, from the very nature of his plumage, could not fly except against the wind, illustrated to them the habit and necessity of approximate virtue in a world of violence and temptation. Supreme pride, and an unconquerable love of freedom, were seen in the quetzal, a native of the tropics, resembling a parrot. It was so constituted that if but one of its feathers was plucked it instantly died. If an attempt was made to cage the strange feathered visitant, it deliberately attempted suicide by pulling out its own feathers, preferring death to captivity. A species of variegated woodpecker, called the carpenter, interested them, for the fidelity and devotion it exhibited. If one were killed, it was rare that its mate did not come and place itself beside the dead body, as if imploring a similar fate.

An argument deduced.

Fidelity and devotion.

The wren was their type and model of content and confidence. Instances were known where young ones that had been disturbed and threatened were found in the nests of robins, by whom they were fed and protected. They did not like the cuckoo, for the incarnate selfishness it displayed. It would deposit its eggs in the nests of other insectivorous birds, not more than one in a nest, leaving the care of the young entirely to the foster parents thus selected. A distinguished poet and close observer of nature was asked why it happened that so many young singing birds were lost for a single young cuckoo. In the first place, he said, the first brood is generally lost; for even if it should happen that the eggs of the singing bird are hatched at the same time with that of the cuckoo, which is very probable, the parents are so much delighted with the larger bird, and show it such fondness, that they think of and feed that alone, whilst their own young are neglected, and vanish from the nest. Besides, the young cuckoo is always greedy and demands as much nourishment as the little insect-eating birds can procure. It is a very long time before it attains its full size and plumage, and before it is capable

The cuckoo.

Always greedy.

of leaving the nest, and soaring to the top of a tree. And even a long time after it has flown it requires to be fed continually, so that the whole summer passes away, while the affectionate foster-parents constantly attend upon their great child, and do not think of a second brood. It is on this account that a single young cuckoo causes the loss of so many other young birds. But they did enjoy the blackbird, for his loquacity and gregariousness. They had studied his language, and understood him when he talked. Their interpretations were very amusing. Nothing delighted them more than to see him bathing in moulting time, and he alike enjoyed the admiration he excited. There were places along the shallow streams where great flocks assembled for that purpose. Half an hour before sunset was a favorite time for the entertainment. Successively and simultaneously they rose out of the water, chattering as they ascended, and shaking out their glittering plumage, they filled the air with myriads of rainbows — reminding observers of the ascent of the great groups of gay butterflies, described by travelers in the tropics, — orange, yellow, white, blue, green, — which, on being disturbed, rise

Do not think of a second brood.

Myriads of rainbows.

from the moist beach of the pools into the air by hundreds and hundreds, forming clouds of variegated colors.

Insects and reptiles of all sorts were objects of interest and study. Nothing pleased the children more than to fasten a little snake in the grass with a forked stick an inch or two behind its head, and on their knees with a good glass to look inspectingly into his interesting face. And the little beauty, they always said, looked into their faces with as much interest as they did into his. Some sensation was created in an electric railway carriage, where there were many passengers, by the escape of a boxful of mountain adders; but the boy soon gathered up his pets without damage or difficulty. The wife of a distinguished naturalist found one morning in one of her slippers a cold, little slimy snake, one of six sent the day before to her scientific husband, and carefully set aside by him for safety under the bed. She screamed, There is a snake in my slipper! The savant leaped from his couch, crying, A snake! Good heavens! Where are the other five? Strange, the people naturally exclaimed with the philosopher, that nature

Insects and Reptiles.

The savant's exclamation.

was never so powerful as in insect life. They were ever ready with striking examples. The white ant could destroy fleets and cities, and the locusts erase a province. And then how beneficent they were! Man would find it difficult to rival their exploits: the bee that gave honey; the worm that gave silk; the cochineal that supplied the brilliant dyes. But infusoria! One saw in a little drop of water on a piece of glass a whole world of insects, of which the largest looked like grasshoppers, the smallest as pins' heads. Some of them were really like grasshoppers, others had the most monstrous shapes, all were tumbling about each other, and the big ones swallowed their smaller neighbors. He saw infusoria in his own blood; it swarmed with eels and cod and all sorts. It was no optical illusion; he saw the forms of the insects and the movements of the different joints; and besides, when he touched the globule with the point of a pin dipped in acid, they at once fled to the other side and died a moment after. The white mould in ink appeared a great forest, with plants, trees, and bushes; the infinite opened before him, and he turned dizzy.

Infusoria.

White mould in ink.

In Universal Sympathy

All this to give some idea of the character and mental resources of the people. They were simple in tastes and philosophic in tendency, and their humanity was broad enough to cover every living substance. This love of life, and perception of conscious existence, brought them in contact and sympathy with every pulsating organism, whether of man, animal, insect, bird, or reptile. The smallest living object was as wonderful to them as the greatest, and commanded their admiration and reverence. Their greatest happiness was in intellectual and moral activity. The possibilities of mental achievement and moral elevation determined their aims and duties. This tendency to universal investigation had not only established the feeling of universal brotherhood, but had opened the way to its possible accomplishment. Nothing seemed small that looked to that end. The utmost that any one could do was only a little — the aggregate was the crown of mortality. Man was less to them than men, but manhood was above the mass, and not to be compounded. That was scrupulously in view and practice throughout all their education and civilization. It had the good effect to fix responsi-

Character and Mental Resources.

The crown of mortality.

bility. Society was not held responsible for conduct, however much it might influence it. The individual was the immortal, and not the multitude. Multitudes might dissolve, as solid bodies, into particles, but individuals, as atoms, were not lost in the dissolution. The utmost estimate was put upon a just and enlightened man, and he was not disparaged nor degraded but by himself. There were limits to fusion with the multitude. Surrender was incompatible with sound growth. Discipline was much, but did not constitute character. Wheels and cogs were not the motive power. Character grew by individual endeavor, and was exalted by worthy aims. Powers were developed and determined by being constantly tested. A thing acquired by the man himself was more than acquisition, it was discovery. The habit of individual effort and investigation made it impossible to knead men into masses. Their intrinsic and indestructible personality occasioned only effervescence and explosion whenever the attempt was made, — which was not often, as a memory of consequences was not quick to die out. The people were very generous in compromise, but not to the extinction of personal rights and obli-

The individual the immortal.

Not to be kneaded into masses.

The Business of Society

gations. Their tolerance was unqualified, but as the principle of give and take qualified it. They gave as they demanded. Impatient of intrusion, they did not intrude. The business of society was to help the individual, not to absorb him. Where every man was a man, that was impossible. There was not anything of which the Sub-Cœlumite was so sensitively jealous as of the undisturbed possession of the ground each one stood upon. His title was of God, and his ownership was not to be questioned. He met his obligations and acknowledged citizenship, but not to the last extremity. There was always a point where compliance would be extinction or slavery. Abreast and arm-in-arm he was willing to move generally, not always. As he respected himself he respected others. He would not tread upon nor be trodden. He granted the large liberty he exacted. By his personal, individual efforts he had become contradistinguished, as every other man had who deserved the name. He had not aimed to be like any other, but to be himself. His study of the ant had made him reverent of him, as of the species. The gifts of individual and associated character appeared in the insect

Help to the individual.

Contradistinguished.

commonwealth as in Sub-Cœlum. They matched as they were known, and were not underestimated as they were perceived in either. Intelligence, in whatever creature, inculcated humanity, as sentient existence inspired reverence. All was of God, for His own wise purposes, and inestimable but by Him. From ant to man was a sweep the Sub-Cœlumite did not pretend to compass. He bowed low, and trusted.

Intelligence inculcated humanity.

Personal independence — born of intelligence, plain living, and individual development — was a marked characteristic of the population. Habits of reflection, self-denial, and just self-estimation, made them poor material for the demagogue and crafty churchman. They could not be trained, at will, to perpetual thoughtless subordination and submission into sects and parties. Not that they resisted coöperation, but that conditions were always changing, and that the point of observation of any one was never long exactly the same. Deference to others did not signify involuntary surrender of themselves. Patriotism made them generous in political action, but not heedless, nor personally irre-

PERSONAL INDEPENDENCE.

Conditions always changing.

Charitable to Others   207

sponsible. Being essentially religious, in all that the word implies, they were charitable to others alike so, and were unfitted for sectarian antagonism. Feeling and judgment, operating together, prevented any rash committal that might be embarrassing or unjust to themselves or others. They did not love power for the sake of it. They respected minorities as much as majorities, because of the possibility of their being in the right, and of the probability of their preponderance upon a slight turn of affairs. They were conservative of necessity, because of their reflection and liberality of judgment. Only conscience brought them to a stand of defiance or aggressiveness. What was wrong was not to be compromised with; but the common weal was always of interest in every heart, and divisions were generally upon modes and processes. It had been many, many years since they had been drawn into a war, and that was intestine, and for personal liberty. A crisis had arisen when a distinction was made between the rights of individuals and classes, not in harmony with fundamental policy or sound morals. Every man was guaranteed freedom: each to enjoy the same rights and be entitled to

Unfitted for sectarian antagonism.

Conservative of necessity.

A war for personal liberty.

the same protection as any other. The conflict had not been possible but for inflamed passions and ambitious leaders. Love of power and love of place, aggravated by material interests, arrayed section against section, and blood of brothers flowed, almost without limit. Mere business questions could not have so bitterly divided them, even at that time; but later, from any cause, such antagonism was impracticable. Bloody warfare was an extremity not to be thought of. Leadership that would commit them to it was impossible. To the point of desperation partisan zeal was not to be excited. Leaders, indeed, were only for a season, and then only because they were indispensable. Organization for a purpose did not pledge continuance for any other. Each movement was independent, and not connected with any scheme of personal ambition or emolument. Men were wiser than sheep, who follow their leader whithersoever he may please to lead them. With what devotedness the woolly hosts adhere to their wether; and rush after him, to speak with the rugged philosopher, through good report and through bad report, were it into safe shelter and green thymy nooks or into

[marginal notes: Bloody warfare not thought of. Men wiser than sheep.]

asphaltic lakes and the jaws of devouring lions. It is worth repeating, that, if you hold a stick before the leader, so that he by necessity leaps in passing you, and then withdraw your stick, the flock will nevertheless all leap as he did, and the thousandth sheep shall be found impetuously vaulting over air, as the first did over an otherwise impassable barrier. The people delighted in this illustration of leadership and blind following. In their amphitheatres they repeated it, again and again, for amusement and instruction. Sensitive to satire, and proud of their personality, the lesson impressed itself upon them in a manner to make them distrustful of unnecessary discipline. When they accepted a leader, it was unavoidable, and not without qualification. Following last year was not a reason why they should do the same this. In consequence, dissolution was as inevitable as organization, and a result of it. The ambitious demagogue and subtle priest did not find them plastic in their dextrous hands. As said, they were thoughtful, individual, self-respecting, responsible human beings — not poor, silly, timid sheep, to be led and herded and butchered by kings and priests and heroes without questioning.

Impetuously vaulting over air.

Not plastic in dextrous hands.

Individuality made them interesting.

This individuality made them interesting. Even the average man was not commonplace from conformity, nor the most inferior servile by submission. While of the mass, they were separable, if not self-separated. They avoided, as said, that general language and general manner which tended to hide all that was peculiar — in other words, whatever was uppermost in their own minds, after their own individual manner.

Every man a new creation.

Every man, in their philosophy, as expressed by the philosopher, was a new creation, could do something best, had some intellectual modes and forms, or a character the general result of all, such as no other in the universe had, which needs made him engaging, and a curious study to every inquisitive mind. They did not look at life as a game of checkers, as reformers are apt to do, where every man has the same fixed powers and the same even line of moves. They regarded life, to use the illustration of another,

Life like a game of chess.

not as a game of checkers, but as a game of chess, where every piece has individual characteristics, where every pawn has a chance to be a queen, where the powers and possibility of each piece change with every move or change of square, influenced by past, present, and future, so that every piece may develop into any other by recognition of the law of inequality that presides over individuality, and each move opens new, divine, and wondrous possibilities. That view of life taught each man, if possible, to put a just estimate upon himself, to live appropriately, and to realize, if practicable, his own ideal. He was made to believe, as was truly said, that his real influence was measured by his treatment of himself; that he must first find the man in himself, if he would inspire manliness; that like begets like the world over. The typical citizen, consequently, stood eminently a man amongst his fellows. Genuineness identified him. He did not want any recognition he did not deserve. If influence or fame came to him it was his desert. It was not asked which side he was on. Though possessing the humility of true learning, his mental enlargement was discerned and appreciated. Better than fame, it had been truly said, was the silent recognition of superior knowledge. It was something to be a superior man in Sub-Cœlum. His rank was that of a citizen of the universe, whose mind, as described, was made to

The law of inequality.

Like begets like.

Something to be a superior man.

be spectator of all, inquisitor of all, and whose philosophy compared with others as astronomy with other sciences; taking post at the centre, and, as from a specular mount, sending sovereign glances to the circumference of things. Serene, above the clouds of passion and contending interests, he preserved, to use the happy language of another, that equipoise of manner which told of an equanimity of life. His stature had been determined by possibility. He had made the most of himself within his power. He had been open and receptive, and had invited understanding from all things and all men. Nothing was too small for his consideration, nor too great for his admiration. There was no challenge of superiority, no apparent consciousness of supremacy. Any one might approach him, but no one could appropriate him. Conspiracies did not disturb him, as from their nature they must fall apart. He did not perceive slights, nor care to comprehend their spirit. Envy was oblique admiration. Because great, he did not contend with smaller men in small things. The platform was not to his taste, however worthy of it. Exhibition of himself was a cheapening of his character. The essen-

Equipoise of manner.

Envy oblique admiration.

tial was occult, and did not care to be made self-conscious. It was for inspiration, and not for display. The causes of things are silent, however tremendous may be their results. He did not exact, being sure of a full measure of whatever was his due. Deserving was fate. Impatience was weakness, and evidence of self-distrust. The courage of his heart was for worthy enterprises, and could not be wasted upon trivialities. He did not hurt his powers by an ignoble use of them. Wings for possible flight into the empyrean were not to be impaired by rude uses. His best faculties were for best work, and were not dissipated upon nothings. He did not care to usurp or invade ground already too well occupied. Room, of all things, was what he most wanted, for growth and development.

Deserving was fate.

While absolute personal freedom was secured to all men, no attempt was ever made to produce social equality; that had been left exclusively to self-regulation. The beautiful and interesting law of diversity in all things had been established from the foundation. Out in the forest, under the spreading tree, looking up at the luxuriant

THE LAW OF DIVERSITY.

foliage, you may not think of the difference between the leaves; but pull down a limb, and spend an hour comparing them; you find, much as they resemble, that no two are precisely alike. Examine the plumage of the owl that you cruelly brought down with your rifle; every feather of his beautiful dress differs from every other; and, what is more remarkable, every fibre of every feather is another feather, still more delicate, differing from every other, all of which together yield to the pressure of your hand like floss silk. No wonder he fell upon the mischievous mole or mouse as noiselessly as the shadow of a cloud. Go down to the seashore; the tide is out; there is an apparent waste of white sand, a dull extent of uniformity; but stretch yourself on the beach, which the innumerable differing waves have beaten to incomparable smoothness, and examine leisurely, with a good glass, a few hundred of the infinite grains which you thought to be the same, and you discover that they differ, that each is differently shaped, each holds the light differently, and, what is more wonderful than all, each appears to be a shell, or part of a shell, which was once the abode of a creature, and a different crea-

Plumage of the owl.

Grains of sand.

ture from every other inhabiting, or that ever inhabited, any other shell of the ocean. Look into the crowded street; the men are all men; they all walk upright; they might wear each other's clothes without serious inconvenience; but could they exchange souls? What professor, exclaimed the philosopher, has ever yet been able to classify the wondrous variety of human character? How very limited as yet the nomenclature! We know there are in our moral dictionary the religious, the irreligious, the virtuous, the vicious, the prudent, the profligate, the liberal, the avaricious, and so on to a few names, but the comprehended varieties under these terms — their mixtures, which, like colors, have no names — their strange complexities and intertwining of virtues and vices, graces and deformities, diversified and mingled, and making individualities — yet of all the myriads of mankind that ever were, not one the same, and scarcely alike: how little way has science gone to their discovery, and to mark their delineation! A few sounds, designated by a few letters, speak all thought, all literature, that ever was or will be. The variety is infinite, and ever creating a new infinite; and there

The wondrous variety of human character.

The myriads of mankind unlike.

is some such mystery in the endless variety of human character. Such endless variety was conspicuously seen in the population of Sub-Cœlum. It was impossible, with their intellectual activity and prevailing disposition to make the most of themselves, that it could be otherwise. Freedom of choice in vocation, avocation, and association only made the natural dissimilarity more apparent. Freedom, freedom, without infringement of the privileges, rights, or liberty of others, was the pride of every Sub-Cœlumite. Fetters, gyves, shackles, were his aversion: he would not wear them. Badges, even, he hated, as compromising his freedom. His sense of liberty was shown in an incident in one of the foreign revolutions, when so many persons of different views assumed the tricolor for protection. One well-known person refused to wear it. A workingman meeting him in the street addressed him: Citizen! why do you not wear the badge of freedom? To which the distinguished person replied that it was to show to the world that he was free! In exact proportion to their happy and complete freedom was their unqualified tolerance and liberality. Intolerance was so utterly absent from the

Endless variety conspicuous.

Reason for not wearing the badge.

spirit and habit of their lives that they did not even comprehend it. Why another should be deprived of what they enjoyed themselves was one of the profound mysteries. A distinguished professor in a foreign university showed a visitor a very pleasing print, entitled, Toleration. A *Toleration.* Roman Catholic priest, a Lutheran divine, a Calvinist minister, a Quaker, a Jew, and a philosopher, were represented sitting round the same table, over which a winged figure hovered in the attitude of protection. For this harmless print the artist was imprisoned, and, having attempted to escape, was sentenced to drag the boats on the banks of the river, with robbers and murderers; and there soon died from exhaustion and exposure. The Christianity of the Sub-Cœlumite had survived all the barbarisms of other forms, and was broad enough to include all differences in one compendious unity — his philosophy and religion cherishing and protecting it, as the figure in the picture.

Behind all their civilization, and apparent in every detail of it, was the healthful habit of occupation. It made men self-dependent, self-sacrificing, intelligent, and happy. THE HEALTHFUL HABIT OF OCCUPATION.

Idleness was disreputable. Homes for the Indolent were not established to elevate it, but to warn against it, and to bring additional shame upon slothfulness and inapplication. The servitude of involuntary labor was a quick corrective of the vice of indolence. Poverty was rare, and not a disgrace, except when no effort was made to escape from it. Wants were few and inexpensive. The necessaries of life were cheap and abundant. The vices were not in the market, being largely eliminated. Great sums formerly paid for them were directed to better uses. Appetites were sound, and did not require costly stimulation. Like the passions, they were largely subservient to reason, but in exceptional cases. Enjoyments were found satisfying in proportion as they were pure. Evil propensities and depraved affections were believed to be perversions wholly out of nature. Observation of the habits of animals was a perpetual lesson in moderation. The beasts that perish were decent compared with gross men. The habit of uprightness kept them in line from the centre of the earth to the top of heaven. Hours of labor being few, occupation in the main was voluntary. Well-applied skill and industry

The vice of indolence.

A lesson in moderation.

easily supplied all that was necessary. All labor was alike honorable. Poverty was not dishonorable in itself, but only where it arose from idleness, intemperance, extravagance, and folly, a maxim of theirs descended from the ancients. There were no drones, as every one did something for a living. Whether with brain or hands, every man was a laborer. Sympathy and fraternity were inevitable; contempt, one of another, impossible. Whether in the garden, the workshop, the senate, or the field, each one was accepted a man, and was expected to walk worthily. The grub in the fresh furrow, and the blackbird that devoured it, were resources for his intellect, as the food his labor brought him was sustenance for his body. As he trod the clods, the earth moved to meet him. Whatever his occupation, when he stepped out under the blue dome, and looked up at the galaxies, he beheld, with the enraptured poet, the Street-lamps of the City of God. His mind his kingdom was, and not the shop or farm, ever and ever. He was less for the morrow than for the everlasting. Leisure, to those who knew rightly how to employ it, they held with the philosopher to be the most beautiful of possessions;

Every one did something for a living.

His mind his kingdom was.

yet without this knowledge it became burdensome and a fate. One must, they said, espouse some pursuit, taking it kindly at heart and with enthusiasm. Fruit he must bear or perish of lassitude and ennui. Leisure to be perfectly enjoyed must be earned — then it is divine. It opens the windows of promise, and receives what it invites, to fullness. Rightly employed, as in Sub-Cœlum, it fills society — to borrow just words — with gentlemen, of inherent self-respect and inherent courtesy; it fills it, also, with ladies, of purest mould and divinest exemplariness. It made the people self-sacrificing, with opportunity. It was a maxim with them, that man is never wrong while he lives for others; that the philosopher who contemplates from the rock is a less noble image than the sailor who struggles with the storm. Recognition or compensation of humane service was not in the least a consideration. The lesson of the Wise Man, in language and spirit, was ever before them: There was a little city, and few men within it; and there came a great king against it, and besieged it, and built great bulwarks against it: Now there was found in it a poor wise man, and he by his wisdom delivered the

Leisure that is divine.

Lesson of the wise man.

city; yet no man remembered that same poor man.

The frequent use of the words probably and perhaps, and their equivalents, was characteristic of the people. It showed that consideration and deliberation were habitual in their speech. Care was taken to impress upon the young the importance of these words. They were printed upon cards and hung upon the walls of schoolrooms. Sentences illustrating their value and correct employment were written on the blackboards. In these ways the difficulty or impossibility of absolute knowledge was stamped upon the growing mind, and the necessity of circumspection in speech impressively enforced. They were taught the importance of habit — in that as in everything. Frequent reiteration fixed in the memory the valuable precept, Choose the course which is best, and habit will make it easy. Truth holding the first place in their system of education, approaches to it were opened and guarded in every practicable manner. Frequent repetition was required to make the pupils accurate, and to impress them with a sense of accountability. Dogmatic statement, from

its very nature, was suspected. It closed every avenue but the one traveled over by him that made it. It also had an element of violence in it that was inimical to just *Disputation.* thinking. Disputation was its life. There is an account of an orator who was wonderfully choleric by nature and indulgence; to one who supped in his company, a man of gentle and sweet conversation, and who, that he might not move him, approved and consented to all that he said; he, impatient that his ill-humor should thus spend itself without aliment: For the love of the gods! contradict me in something, said he, that we may be two! When thinkers met together to think, or dilate, they did not, so to speak, answer one another; they permitted to thought the utmost freedom, consistent with just intellectual hospitality, and did not antagonize it; they might differ from it, but not by direct reference. Thought stimulated but did not provoke. Disputation was out of the question in independent thinking. While each one was free to express himself, a like liberty was *Dogmatic* not denied to any other. Dogmatic dis-
discussion. cussion was not consistent with their conception of intellectual growth. Where each one knew a little, and no one pre-

sumed to know all, the way to a fair understanding was not difficult. Feuds were discouraged in every possible way. Hard names were not given to men and things, certain of their reaction, as of their injustice. The habit was to say the most favorable things of others, and to avoid detraction. If a harmful thing was idly or viciously said of a neighbor, some one present was sure to make a note of it. If not apologized for and withdrawn, accountability for it was fixed. Dangerous gossip in this way was largely prevented. Truth and falsehood were discriminated. Visiting faults and sins upon those innocent of them was not a fashion of general adoption. Their religion was against it as well as their habit and philosophy. They looked to their own conduct, rather than to their neighbors'; for it they were accountable, and not for theirs. Ever present with them, and not to be forgotten, was their profound sense of personal responsibility — the foundation and superstructure of their ethics and religion. All of which promoted good neighborship and inspired security. The man, they said, who delights in giving you full credit for every excellence you possess, rather

The habit to say favorable things.

Accountable for their own conduct.

A treasure. than in belittling you by an exaggeration of your foibles, is a treasure; and the protection you feel in the neighborhood of such a man, law could not give you. He shuts your gate, he protects your child, he guards your reputation; he does the fair and generous thing. If men were weighed and not counted, such an one would overbalance many of poorer material. A wise man, having a farm to sell, bid the crier proclaim also that it had a good neighbor.

THE SOCIAL CONSCIENCE. It was another of their maxims, that misunderstandings and neglect occasion more mischief in the world than even malice and wickedness, and they looked to them especially. Only the very few indeed, by what has been called the alchemy of private malice, concocted a subtle poison from the ordinary contacts of life. For the fun of the thing, not for the mischief of it, the world there, as everywhere, prattled on. Sometimes it was cruel; but it was the cruelty of the thoughtless boy. It did not much concern itself, for the time being, about justice or injustice. To the sources it did not much care to go if it could. It preferred to see with its eyes rather than with its head, — by its senses rather than by its reason. It saw outwardly, and talked for recreation — irresponsibly, too often, and without reflection. When it criticised or ridiculed, it did not always consider that the best continually blunder and stumble, and only learned to keep their feet by falling. Morally as well as physically. If an invisible knocking machine tapped each one on the head the instant and every time he meant evil or thought wrong, what a getting up there would be! What a scene the street would present! To the church or the market the same. The world laughs; — with us, and then at us. Careless words sometimes left their sting, and rankled long after they were uttered. Repeated, the wound was less curable. Thy friend has a friend, and thy friend's friend has a friend; be discreet; was a saying they did not always carry in their minds. The inward wounds that are given by the inconsiderate insults of wit, they did not always wisely remember are as dangerous as those given by oppression to inferiors; as long in healing, and perhaps never forgiven. Particular pains were taken to impress these truths upon the less reflective. They were taught the danger of idle personalities, and that the mischiefs

The best blunder and stumble.

Inconsiderate insults of wit.

they created were sure to be permanent if not soon corrected. A habit of often reviewing their social relations was urged, and pretty generally adopted. Explanation was promptly made whenever it was thought just and merited. If the slightest cloud was discovered on an acquaintance's face upon meeting him, time was not lost in removing it. If avoidance was perceptible in the conduct of any one, the reason of it was sought, and good relations were restored. The social conscience was quickened and enlightened by these good offices. While it was not possible, with the utmost circumspection, to altogether prevent misunderstandings, it was found easy to correct them by going a little more than half way towards it. The consciousness of possible offense was enough to prompt explanation and apology. While words and circumstances were remembered, and not aggravated or perverted by brooding, candor and truthfulness were sure to make all plain and satisfactory. Malice was thwarted by anticipation and prevention, and memory was not even disturbed by the remembrance of misconception or difference. A better understanding was established, and the friendship temporarily

Kindness to Children

lost was made permanent. Neglects were atoned for by greater consideration and kindness. Affection was fed by tenderness, and starved hearts restored by bounteous sympathy. Ill-treatment of children was one of the gravest of social offenses. It was considered a mean and cowardly iniquity. One of the distinguishing marks of a thorough gentleman was his considerate kindness to children. Their favorite novelist had said — a favorite on account of his searching, sympathetic, profound humanity — that in the little world in which children have their existence, there is nothing so finely perceived and so finely felt as injustice. It may be only small injustice that the child can be exposed to; but the child is small, and its world is small, and its rocking-horse stands as many hands high, according to scale, as a big-boned coursing hunter.

A grave social offense.

Amusements were simple — as far as possible educational and hygienic — and adapted to the multitude. The tone of their theatres was generally elevated — in no sense degrading. Comedy and tragedy of the highest order were preferred. Stage dress was limited to decency. Representa-

Amusements.

tions that would occasion a blush the public taste prohibited. Applause was judicious, and never clamorous. Doors were closed before the performance commenced.

The theatre. Disturbance from going in and out was not permitted. People went to see the play, and not to display themselves. Showy dress was considered vulgar — refined people avoided it. At the opera, greater freedom was indulged; the audience being a larger part of the entertainment. Eyes were feasted at the same time that minds and tastes were gratified. As before said, the people most delighted in oratorio, and their dress and behavior were much the same as at the theatre. Too elaborate adornment made them self-conscious, and limited their enjoyment of the higher, bet-

The circus. ter music. The circus was more generally popular than any other entertainment. Its character brought together great audiences — appealing especially to the senses. The masses of humanity, comfortably seated and happy, were a great spectacle. Twenty thousand was not an unusual audience. Physical education was inspired by the amphitheatre, and added interest was given to the gymnasium. Pedestrianism was a favorite amusement and exercise of the people. It taught grace, and gave vigor and health to the constitution. It stirred the mind, whetted the appetite, and drove away melancholy. So common was the healthful diversion that no able-bodied person thought of spending a day without a long walk. Their beautiful roads were most inviting to pedestrians. In favorable weather, walkers were never out of view. Women as well as men enjoyed the pastime. The grace and beauty of their movements were a perpetual charm. Springs of sweet water were at convenient distances on the highways, affording delightful resting-places. Manly men and womanly women exchanged courtesies. Bright eyes and rosy cheeks and musical voices animated these natural and accidental meetings. Cupid was close about, and Hymen not far off, and nobody could guess what a morning would bring forth. Dancing, of course, was a chosen amusement; but it was scrupulously limited and guarded. Public balls, where anybody might be admitted for the money, were not tolerated — even by the most inferior of the population. The universal self-respect tabooed all such degradation. Pyrotechnic displays were common, especially on anni-

Their beautiful roads.

Hymen not far off.

versaries and other popular occasions. Great crowds assembled to witness them. Perfect order prevailed in these street assemblages. Not a word was spoken that was unfit to be heard, nor a glance or movement ventured that could offend.

Kite flying. Kite flying was universal; it seemed to be the one outdoor amusement that everybody loved. Old and young participated in it. Their kites were mechanical and scientific wonders. They were ingeniously constructed, and rose as naturally and gracefully as birds. Some of the designs were very beautiful and suggestive. For hours and hours together all ages amused themselves with all manner of aerial contrivances. Spelling-contests had long been kept up, and the people never wearied of attending them. A high premium was put upon perfect spelling. It was felt to be a shame not to be able to spell any word in common use amongst intelligent people. Rewards were paid to perfect spellers, and distinction was conferred upon them.

Reading. Reading, also, was a public exercise, and was of great service in general education. As so great a part of their pleasure and instruction came through reading, the greatest effort was made to improve themselves

The True Standard 231

in it. In the book of Nehemiah they found the true standard of reading aloud — how Ezra, the learned and pious priest, and the Levites, read to the people the law of Moses: they read in the book, in the law of God, distinctly, and gave the sense, and caused the people to understand the reading. The rule of Ezra and the priests was the rule adopted throughout the Commonwealth, which, by its very nature, discouraged anything like elocution. It produced a multitude of good oral readers, who penetrated the words of the printed page, perceived their sense, and participated their feeling, and were able, unconsciously, to interpret, reveal, and enkindle them in the reading. Chemical experiments were constantly made for the edification and amusement of the people. They were taught the chemical elements, and all their known offices in nature. Such practical instruction helped them in outdoor observation, which, at last, was their best resource and entertainment. Some pains have already been taken to show the reader how the population were interested in everything that existed — from creature to man, from atom to sun, from sun to universe. Their habits of observation

The rule of Ezra.

The chemical elements

made their minds acute, and their close sympathy with nature exalted their souls. To repeat, they were with God in His works. Each season produced its wonders. To see a noble forest, they said, wreathed in icy gems, was one of the transcendent glories of creation. You looked through long arcades of iridescent light, and the vision had an awful majesty, compared with which the most brilliant cathedral windows paled their ineffectual fires. It was the crystal palace of Jehovah.

With God in His works.

In the province of Kadoe is the great temple of Boro-bodo, described by travelers in the tropics. It is built upon a small hill, and consists of a central dome and seven ranges of terraced walls covering the slope of the hill and forming open galleries each below the other, and communicating by steps and gateways. The central dome is fifty feet in diameter; around it is a triple circle of seventy-two towers, and the whole building is six hundred and twenty feet square, and about one hundred feet high. In the terrace walls are niches containing cross-legged figures, larger than life, to the number of about four hundred, and both sides of all the

DRAWING, PAINTING, AND SCULPTURE.

terrace walls are covered with bas-reliefs crowded with figures, and carved in hard stone; and which must, altogether, occupy an extent of nearly three miles! The amount of human labor and skill expended on the Great Pyramid sinks into insignificance when compared with that required to complete this sculptured hill-temple in the interior of a tropical island. A philosopher told a story of one of the lords of session in his country, a strange, rough, gruff judge, who was in the habit of taking sketches of people in court with a pen and ink. One day he asked the usher, Who's that man yonder? That's the plaintiff, my lord, was the answer. Oh, he's the plaintiff, is he? he's a queer-looking fellow; the Court will decide against him and see how he'll look! History goes not back to the time when art in many of its diversified forms was not practiced. In Sub-Cœlum the taste for it was universal, and great progress was made in its cultivation. The artist's eye and habit had been quickened and strengthened by the generous system of instruction. The principles and practice of drawing were carried into all their schools and intelligently taught. Perhaps one pupil in fifty discov-

Philosopher's story.

The artistic taste universal.

ered ability, and was encouraged; if one in ten thousand showed genius, there was hope; but the multitude was benefited. Taste was cultivated if nothing more. Adepts in drawing were not uncommon. The little books in side-pockets contained many admirable sketches. They revealed the searching observation of faces that the judge in the story exhibited. Thumb-nails were shaped to use in sketching. A very small card in the artist's hand would receive and retain necessary outlines. In public places there were conveniences for posting anonymous and other original drawings. Very acute many of them were, and taught as the most logical discourses could not. A little picture would illumine a public question. Caricature was of course indulged, but not dangerously nor licentiously. Private character, unless connected with the public in a way to occasion mischief, was sacred to it. Women also, whatever the folly to be exposed, were never subjects of ridicule or open attack. There were limits that the public had severely prescribed, and they were rarely transcended. The artist who misused his pencil or brush became odious. He was not tolerated. If incorrigible he

Searching observation of faces.

Private character sacred.

Human Nature Exalted 235

was locked up. The public taste ran to the virtues, and delighted to see them represented. Infinitely were they exhibited, in pencil and in color. Human nature was constantly being exalted by these representations. Sculpture, especially, employed itself in embodying the highest qualities and achievements of manhood and womanhood. Martyrs to reason, to humanity, and to personal freedom, were the favorite subjects of superior genius. Heads and figures of Socrates, of Jesus Christ, and of John Brown, were to be seen in public places. The brow of the first appeared the home of intellect; the face of the second shone with a supernatural light; the front of the third was rugged, like the brow of Hercules. These representations, idealizations, realizations, were instructive and elevating according to the mood or extremity of the beholder. An intellect in shadow, ill-recognized and unrequited for the time being, gained courage in contemplating a head of the brave philosopher; a poor fellow, feeling himself oppressed, recovered hope as he paused before an ideal representation of his hero; a woman, in anguish, uncovered before a figure of the immaculate Saviour, and cast an upward

How sculpture was employed.

Instructive and elevating.

look of adoration that no eye witnessed without sympathy. Blessèd be art, they said in their hearts, that lifts us up when we are cast down; that puts a hope into discouraged souls; that exalts wretchedness to a place in the bosom of Deity. There was not any person or place that did not feel the pervading influence. Homes were adorned by it, and flooded with a healthy moral atmosphere. Not one but had ideals of virtue that were perpetually teaching. Shame covered the face of wrong in their pure presence. Sculptural manipulation of clay was one of the common amusements. The expert would take in his hand a portion of kneaded earth, and exhibit the passions and emotions one after another, as they were asked for. Grief would drop a tear over the thumb-nail, and Santorini's laughing-muscle show itself in the face. Horrible were some of the faces made, and lovely were others as genius could make them. Draughtsmen, in goodly number, were alike capable in their department. On the blackboard, or other suitable drawing surface, they gave to observers whatever expression or outline they requested. Animals were drawn with human-like faces, and men with the faces of animals. Wings were transferred from birds to reptiles. There was no limit put upon these diversions except by time. Audiences broke up with abated respiration.

Books they had in abundance — too great abundance, they constantly felt. With all their weeding, the number was not lessened. They were not ambitious of great libraries, quality being preferred to quantity. Their aim was to preserve only the best. They realized that minds, like some seed-plants, delight in sporting; there is great variety in thinking, but the few great ideas remain the same. They are constantly reappearing in all ages and in all literatures, modified by new circumstances and new uses; though in new dresses, they are still the old originals Like the virtues, they have great and endless services to perform in this world. Now they appear in philosophy, now in fiction; the moralist uses them, and the buffoon; dissociate them, analyze them, strip them of their innumerable dresses, and they are recognized and identified — the same from the foundation and forever. If a discriminating general reader for forty years had noted

Not ambitious of great libraries.

The few great ideas.

The same from the foundation.

their continual reappearance in the tons of books he had perused upon all subjects, he would be astonished at their varied and multiplied uses. Thinkers he would perhaps find more numerous than thoughts; yet of the former how few. The original thought of one age diffuses itself through the next, and expires in commonplace — to be born again when occasion necessitates and God wills. At each birth it is a new creation — to the brain it springs from and to the creatures it is to enlighten and serve. If the writer or speaker could know how often it has done even hack-service in the ages before him, he would repentantly blot it out, or choke in its utterance. In the unpleasant discovery, that indispensable and inspiring quality, self-conceit, would suffer a wound beyond healing. In literature, as a rule, the oldest books were preferred; in science the newest. The classic, they said, was always modern. Simplicity they considered, with the critic, the last attainment of progressive literature: as men are very long afraid of being natural, from the dread of being taken for ordinary. They accepted the definition of literature to be the written thoughts and feelings of intelligent men and women arranged in a way

Thinkers more numerous than thoughts.

Simplicity the last attainment.

to give pleasure to the reader. Pleasure could not be had where there was affectation, and where meaning had to be groped for. Perspicuity was an essentiality. The miserable habit of some biographers of searching out the weaknesses of authors with their audacious dark-lanterns, was not in favor in Sub-Cœlum. Men had a right, they said, to be themselves, if they were authors: and they were not to be called hypocrites if their thoughts and conduct did not always agree. It was from this sublime inevitable simulation of literature, they said and repeated, that the world gets its lay working ideal perpetually renewed. As yet, a human creature can only sometimes be quite good in the still act of writing. By a happy error those who do not write mix up the man and the author, where the difference is not forced on them, and thinking there are beings so much better than the common, they try fitfully to live after the style of books. If the illusion should be destroyed, and it ever came to be universally known that literature is intentional only, that the writers of these high judgments, exact reflections, beautiful flights of sentiment, are in act simply as other men, how is the great bulk to be stung into try-

Thoughts and conduct.

Literature intentional only.

Metaphysics. ing after progress? Metaphysics, having long ceased to be considered a science, books on the general subject were scarce; they had mouldered away, or been consigned to the paper-makers. The same *Political economy.* judgment of political economy had reduced the books upon that subject to a few. The political economist, they said, looked upon men too much as machines, and his system, they thought, contained too many conflicting calculations and theories to be useful. Masterpieces of authors were scrupulously treasured; indeed it was their rule, with voluminous writers, to preserve only their greatest achievements. Those books that the ages had passed upon were accepted as indubitably worthy. They believed with *No luck in literary reputation.* one of the greatest that there was no luck in literary reputation. They who make up the final verdict upon every book are not the partial and noisy readers of the hour when it appears; but a court as of angels, a public not to be bribed, not to be entreated, and not to be overawed, decides upon every man's title to fame. Only those books come down which deserve to last.

THE PRESS. The tone of the press was such as might

The Antidote

be expected from the character and intelligence of the people. It was moderate, but wholly and habitually free. As well said, a press is mischievous only where it is partially and irregularly so. Just as a draught gives you a cold, while even a storm in the open air is innocuous. If the press were free for a fortnight only in every year there would be an annual revolution. Its duty, as defined by a distinguished member, was to make war upon Privilege — to see that a ruling class was not formed in the State, to reduce the functions of officials, to eliminate from the popular apprehension the illusions of political superstitions. It adopted as a maxim, The less government the better; the fewer laws and the less confided power. The antidote to the abuse of formal government, they said, was the influence of private character, the growth of the individual. Journalism, adopting the language of a critic, was pitched on a low key, and set about on the ordinary tone of a familiar letter or conversation; as that from which there was little hazard of falling, even in moments of negligence, and from which any rise that could be effected must always be easy and conspicuous. A man fully possessed of

Its duty.

Pitched on a low key.

his subject, and confident of his cause, may almost always write with vigor and effect, if he can get over the temptation of writing finely, and really confine himself to the strong and clear exposition of the matter he has to bring forward. Accuracy and definiteness were of the first importance in their journalism. Violence was suspected — even strong language — except in rarest cases. Italics were not used, as every word was expected to italicize itself. Intelligence was discriminated and severely sifted. News was not anything that might be invented, embellished, or perverted. It was the rule to publish only what was literally true. News gatherers were instructed to be direct and concise. A column about a trifle was not acceptable. Ability in condensation was preferred before facility or felicity. While personal items were sought and desired, great care was taken to print only such as were respectful and creditable. Journalism generally had adopted as a motto and rule of conduct a sentence from a famous writer: Private vices, however detestable, have not dignity sufficient to attract the censure of the press, unless they are united with the power of doing some signal mis-

Accuracy and definiteness.

Motto and rule of conduct.

chief to the community. Objectionable matter, from its nature, found a place in The Chronicle of Perdition, a journal that, in spite of public opinion, found a sufficiency of readers to support it. Alas! there were people, even in Sub-Cœlum, with prurient tastes and appetites, who de- *Prurient tastes and appetites.* lighted in recitals of evil and gross criminality. A journal of general circulation was called Information for the People. It was crowded with condensed facts upon all sorts of subjects, and formed a literature of its own. It was intelligently indexed, and had grown into many large volumes. It was a mine of information, that was constantly consulted by all classes. But the most popular of all their journals bore the significant title of Confidential Letters to the Public. Each number of it contained a hundred or more free communications, from as many persons and places, upon a great variety of subjects. It was sometimes called The National Barometer. It *The National Barometer.* indicated the matters upon which the population were generally thinking, and especially those about which they were most uneasy. Questions were discussed, but not in an elaborate manner. Space was too valuable to permit the inundating method to any. Grievances of all sorts were acutely and forcibly presented.

Functionaries consulted it. Functionaries, especially, consulted the suggestive journal for cues, and assemblymen referred to it as authority. No worthy subject, of social or political interest, escaped investigation in Confidental Letters. Communications were anonymous, but the names of authors were registered, and produced, if in extremity they were called for. It was not possible for any intelligent citizen to avoid being interested in its contents. It determined for him the average judgment upon current topics; it

It gauged apprehension and anxiety. put his finger upon the public pulse; it gauged apprehension and anxiety with approximate accuracy. Nothing unhealthfully stimulating, as a rule, was found in their newspapers. Sensation was not in favor; truth and decency were elevated above everything. They were not ambitious of the picturesque or startling in their annals; on the contrary, they preferred the commonplace and tiresome, as more significant of contentment and prosperity.

RESULTS OF EVOLUTION. In the evolutionary processes of this peculiar civilization some unexpected changes had resulted. The dogs did not bark noisily, as had been their wont; the moon, even, did not disturb them. They contemplated Luna, but without demonstration. The cats, likewise, were considerate of the peace of neighborhoods. Men, many of them, changed places with women, and became essentially domestic. Household duties, in a great degree, had passed into their hands. They discovered a fondness for them, as to the other sex they became distasteful. In well-to-do households every department but the nursery was surrendered to them. They were strong, and could lift, and climb, and stoop, without difficulty or detriment. The kitchen, especially, was their domain. Cooking, as before observed, was a very high art in Sub-Cœlum. Learning had been devoted to its development. Chemistry, particularly, had been ransacked, and its mysteries applied extensively. Kitchens were laboratories and museums. Contrivances for everything had been invented and appropriated. Cook books had grown to the proportions of cyclopædias. As the word servant was obsolete, and never used throughout the Commonwealth, the profession of cook was as respectable as any

Men became domestic.

The kitchen their domain.

Cook books cyclopædias.

other; indeed, a master in the kitchen ranked with scholars and scientists. To his genius they attributed much that was best in their life and achievements. In their profound study of body and mind — of their dependence and interdependence — how astonishingly morals depended upon stomach — the necessity of good cooking was appreciated, and the art elevated. Soups were in such variety that every want of appetite and emotion was provided for. A dinner for the gymnast and a dinner for the poet were as different as any two things of a kind could be. The resources and gamut of the emotions had been studied as profoundly as the possibilities and power of the muscles. Training for anything remarkable was largely through the wisdom and manipulations of the kitchen. Eating was determined by occupation. The orator prepared himself for highest flights by days of discriminate living. The clergyman, to impress his hearers, was conscientious about his breakfasts. It was not thought possible for a judge to be considerately just without judicious and temperate diet. The actor, especially, was indebted to the cook for his reputation. The green-room and the

Necessity of good cooking.

Eating determined by occupation.

Enjoyment Inevitable 247

kitchen were inseparable to him. Dinners in well-ordered households were inspirations, the cook having eaten appropriately to achieve them. The dishes were so wisely various, so divinely cooked, and so perfectly served, that enjoyment from them was inevitable. Conversation was in keeping, and men and women regarded themselves as worthy of the perpetuation they hoped for. The cook commanded better wages than the senator. Anybody, after a fashion, might perform the functions of the latter; the skill of the former was exceptional and essential. The perfect cook was a desideratum in that high civilization. At banquets, the chef appeared at the end of the entertainment and received his just homage. Pledges were drank, and wine poured out in honor. Guests rose, and bowed low, as their genius and benefactor passed out. Grades there were, of course, in the profession — in ability and dignity; but there was pride in it throughout, and every member of it studied to attain the utmost excellence. Households were happier with male cooks; the women preferred them, and treated them as gentlemen. Servant or scullion was not thought of in the pleasant relation.

Conversation in keeping.

Pride of profession.

The high estimate put upon woman was evidence of incomparable advancement. Feminineness, whether in virginity or maternity, was exalted. No man forgot to pay reverence to the sex of his mother, his wife, or his sweetheart. Adoration of the Virgin Mother was its apotheosis. Oh! exclaimed the humanist, if the loving, closed heart of a good woman should open before a man, how much controlled tenderness, how many veiled sacrifices and dumb virtues would he see reposing therein. As far as possible woman was emancipated from menial duties. The offices of motherhood, especially, were not infringed by avoidable domestic drudgery. She was left free to devote herself to the care and development of her children, and to the enjoyment of such society as would supply the want occasioned by continually descending and imparting. All suitable occupations were thrown open to women, and some of them they monopolized. It was found that they made the best physicians — especially for children and women. Their delicacy and courage made them superior surgeons. Their fingers manipulated in a manner impossible to men's. In cases of confinement they were pre-

Remarkable Intuitions 249

ferred, without exception. Women in that crisis reasoned, as reported, and were listened to deferentially. They said frankly, if pressed in so delicate a matter, that all their strength, in the act of violent exertion, consisted in the liberty of the exertion, and that this liberty was as nothing if a man was in the room. From this cause, at every moment, hesitation resulted, and contradictory movements. They exerted and they restrained themselves. You will say, says the wise reporter, that they are in the wrong, that they should be at ease, should, in such a crisis, forget their superstitions of shame and fear, the little annoyances which so humiliate them. But, however this may be, such they are; as such they must be treated. And he who, to save them, will put them in such peril, is certainly unwise. Male physicians, therefore, in such cases, were seldom or never called. In determining the causes of disease, the medical knowledge of women was supplemented by their remarkable intuitions — a very high order of wisdom. As such they were recognized and employed in many important offices. As moral police they kept guard over society. The invisible was duly

A delicate matter.

A high order of wisdom.

rated — nothing escaped their unerring ken. Mysterious and inexplicable, they were nevertheless authority. Judges consulted them in difficult cases. Testimony, contradictory and involved, was analyzed and made perspicuous. Motives were revealed marvelously. The oblique was direct to them. These intuitions were particularly infallible when the conduct of females was in question; for women knew women in Sub-Cœlum. Their knowledge and instincts, so applied, appeared omniscient. Indications unseen and unknown to men were apparent and unmistakable to women. Signs of concealment were as conspicuous as those of unquestioned frankness. Good women were known and read by all; happily there were few in Sub-Cœlum that were not good. Their superior nature was acknowledged and appreciated by all men. It enlightened society and elevated it. Better standards of conduct were set up. Encouragement was given to well-directed effort. Pure and enlightened womanhood was the ripe fruit and governing influence of civilization. It pitched thought and enthusiasm. It adorned whatever it touched. It stimulated charity. It led in religion. The

When infallible.

Pure and enlightened womanhood.

beauty of all things was heightened by it. It was the medium in which all men lived, moved, hoped, and worshiped. The flowers grew better in its atmosphere; the birds sang sweeter; fruits were more deliciously flavored; supernatural rainbows, such as they had, were its typical aureola.

> Her brow
> A wreath reflecting of eternal beams.

Government was largely supported by taxes upon incomes and upon heads, and by a generous system of licenses and annuities. Rich people, being able, were also willing to bear the greater part of the public burdens. It was a privilege they esteemed and were proud of. Estates did not grow enormously. Great possessions were not thought good for the possessors or for the public. They were apt to create distinctions not in agreement with the general system of society and government. The utmost practicable equality was the universal aim. Money was especially appreciated for the leisure it gave to do what was preferable to making it. As repeatedly said before, the ambition of every one was to make the most of himself — to gather resources and treasures that would

How Government was supported.

What money was especially appreciated for.

not fade — that would make him a man in whatever condition or state he might be placed. Believing that this life was only preparatory to a better, every effort was made to develop themselves worthily, and everything not necessary to that was an incumbrance. It was the rarest thing that any one thought money anything in itself. The small tax placed upon every head produced a large aggregate, and it was cheerfully paid. It stimulated patriotism. Every one had a money interest in his Government, and was a supporter of it. When he walked out on one of the beautiful roads, it was his as much as anybody's. When he plucked fruit from the endless orchards, it was from his own trees, that his own money had assisted to plant. The vast and perfect system of schools, by which his children were educated, was not a charity in his eyes, as he and every other inhabitant had helped to establish and support it. The citizen who would withhold his pittance was at heart guilty of incivism. Privileges, in the form of licenses, were liberally and cheerfully paid for. Special rights included special immunities that were inviolable. They were worth more than they cost, and were estimated accord-

Money nothing in itself.

At heart guilty of incivism.

System of Annuities 253

ingly. They included also honor and responsibility. If an individual exceeded his purchased privilege, he was guilty of a breach of trust, and was severely punished. Betrayal was one of the high moral and penal offenses. The system of annuities, as before said, was considerately provident and generous to the people, and was a great convenience. For a sum of money given to the Government, the giver received quarterly a liberal per centum during his lifetime — the amount, of course, being determined by the longevity tables. To scholars and to old people it was a great accommodation. Their savings were turned over to the Commonwealth, and they were supported from them without risk or anxiety. Scholarship was free to pursue its investigations, and old age reposed in the security of independence. In such cases death was not made interesting by possible inheritance. Indeed, it was not thought good that property should descend. Every man, according to their theory, was an accretion, an incarnation; was just what he was naturally, and what he had gathered and assimilated. His personality represented his earnings as well as his attainments. No genuine man wanted what

A liberal per centum during his lifetime.

Every man an incarnation.

he did not earn. It was common for the prosperous to place annuities upon the old and helpless of their kindred, to relieve them of the humiliation and discomfort of dependence. Many rich people, before their deaths, gave away, in this manner, about all that they had, to the eminently needy and worthy. They perpetuated themselves by their good acts, leaving nothing to be wasted in dissipation and indolence. This well-devised system of annuities was not only a pecuniary resource to the Government; it strengthened it also in the affections and interests of the people. Helpful essentially, its judicious and fostering protection was affectionately remembered.

Liberality of rich people.

The machinery of politics, in the sense of office-getting and office-holding, was not strained. Terms of office were short, and elections, of course, frequent. Salaries were small and therefore not greatly desired. Judges alone were elected for long periods, and were paid good salaries. Persons who sought office persistently were mistrusted; desire for place was therefore cautiously and modestly exhibited. Those most worthy were sought

THE MACHINERY OF POLITICS.

by the public. Voluntary preference was gratifying, but place-holding was not considered especially honorable. The man who was fit for a place was not more of a man by occupying it. Merit was in the man himself, and was not increased by recognition. The best men did not hold place at all, except in extremity. Crises sometimes occurred when their services were demanded. In such cases it was manifest surrender, and not for personal advancement or emolument. Titles, though permitted, were not encouraged, and were not often bestowed. They were extraneous, and did not belong to the man; the intrinsic was his personality. Society was filled with men who had been governors and the like, whose rank had not been increased by their temporary eminence. Occasionally one presumed upon it, and arrogated importance in consequence; but the average wisdom and common sense soon relieved him of his conceit and put him back in his place. The airs of a pretended favorite were soon perceived and corrected. A man might be a favorite, indeed, until he assumed to be, when he was not. Public favoritism was fickle and qualified. Gifts were scattered, but lim-

Merit not increased by recognition.

Airs of a pretended favorite.

ited to the public weal. The privileges secured to each were not incompatible with the rights of any. The citizen was elevated by his own worthiness, rather than by factitious assistance. Opportunity was given to all, advantages to none. Every man had an equal chance to make himself what he would. It was not possible to organize men permanently into parties; self-respect and personality forbade. Demagogues were sometimes listened to in times of extremity, but were soon overwhelmed. The ready attention given to ambitious factionists beguiles them to ruin. If the public ear can be easily had, why not its strong right hand, with a dagger in it? Thousands may be got to subscribe a compact of defiance to authority, and the leaders in the scheme of treason may be confident of its success. The roll of names may attain an immeasurable length, and the time for violence arrive. The signal agreed upon, and perfectly understood, is given, when the whole devilish plot appears a failure to its inventors. Those enrolled to participate in the parricidal crime expose and identify their leaders, join in exultation at their disgrace and ruin, and a purer patriotism

Opportunity to all.

The devilish plot a failure.

Objects of Amusement 257

is established. Desperate disorganizers misinterpret public impatience. Their own hearts corrupted, and bent upon disruption and revolution, they assume as much perfidy and baseness in those who listen to and seem to sympathize with them. Popular discontent cannot easily be organized into revolt. An attempt to so organize it, while a particle of patriotism, gratitude, or hope remains, will only quicken a remembrance of benefits, and warm the common heart to a more fervid attachment. Once put upon its guard, no temptation could seduce it. It had been a great while since the people of Sub-Cœlum had been seriously disturbed by demagogues. The few specimens they possessed were generally harmless, and were objects of public amusement. Society was too intelligent, upright, and individual to be long influenced by them. Election day was not more exciting than any other. The utmost independence was secured to the voter, and any infringement of it was rigidly punished. The public conscience at ease, there was little, if any, likelihood of disturbance. Evils were slight, and easily corrected. Clamor was impossible where the people were contented and

Not easily organized into revolt.

The utmost independence secured.

happy. Showy and expensive inaugurations were not in fashion. They were considered vulgar and barbaric. If any demonstration was made it was at the end of a term of office, where the service had been worthily distinguished. Even the chief magistrate quietly subscribed his oath of office, and entered upon its duties without flourish, ostentation, or self-gratulation, modestly impressed with its responsibilities.

Vulgar and barbaric.

Though the people of Sub-Cœlum, as a rule, were good, — good as goodness is qualified and limited by human nature, — they made no pretensions to sanctity; though religious, they were not professors of religion; though Christians, they did not wear badges of piety. There was nothing in the way of dress, language, or manner to advertise super-excellence. Goodness was a personal matter with each one, and was only to be known by character and conduct. Piety was in the life. Genuineness was the standard. Consciousness of imperfection taught them humility. Acutely observing and reflective, they saw God in everything, and were reverent; perceiving the universal dependence, they felt the re-

Essential excellence of the people.

Genuineness the standard.

sponsibility of existence. They truly believed and realized that here we begin to be what we are to be ever. They conscientiously and persistently sought the good and avoided the evil. They carefully guarded themselves against whatever must perish with the body, and ardently cultivated all which must survive it. Happiness was not sought in its transient forms. Life was appreciated by its resultant uses. The duty of the hour was the duty of all time. The good inhered. The present was realized as the period of growth and achievement; and, having something to do worth doing, they needed all the time they had to do it well. The duties of the day faithfully discharged, they did not much concern themselves about the morrow. The morrow was so far provided for that it was anticipated and made easy if it came. Refinement and intelligence and excellence resulted from fidelity to duty, and a happiness was established as serene as it was unconscious. Living and acting, and getting the pleasure and good of life out of each day of it, they enjoyed a foretaste of fruition and perpetuity. They reverenced the life and teachings of Christ for their purity and humanity more than for any

Sought the good and avoided the evil.

The present the period of growth and achievement.

Enjoyed a foretaste of fruition and perpetuity.

dogmas of theology that might appear to be taught in them. They did not understand Christianity to be for the super-terrestrial, to whom sin is known only by wisdom. They understood it to be for men, needing it, and proved its adaptability by accepting it — its practicableness by practicing it. Their Christianity was encouraging, in that it did not require absolute imitation of, but some slight approximation to, the Founder. A religion that was discouraging to hope was a poor religion for men; and a religion that required of them the impossible was such. For some it might be easy to be good — very good — as they understood goodness; for others it was nearly impossible to be good at all according to ideal and exclusive standards. To the former it might seem easy to believe that Christ should be imitated; to the latter it seemed to be only possible that He could be approximated. He was the Great Exemplar, the Divine, to be approached, and only approached, as nearly as possible, by the creature. Now and then, it might be, a man was born into the world in whom were all the virtues so admirably mixed that it was possible for him to approach very near to the Divine

Their Christianity encouraging.

The Great Exemplar.

The Mighty Difference

Founder — so near as almost to touch the hem of His garment; the many, however, were unable to approach so near by a very great way; while the multitudes were so far off that, instead of seeing the light of His countenance, they only saw the reflection of it as it appeared faintly, very faintly, in the happy few, very few, alas! who were able to approach near enough to feel a little the direct rays of the Divine Effulgence. After a poor creature had done all that it was possible for him to do, it was discouraging to be told that he had not done enough; that after he had done all that it was possible for him to do, he should be lost. He knew himself what he could do and what he could not do; and found himself unable to accept a form of faith which offered rewards for the impracticable and impossible only. If the gate of Paradise was to remain shut against him, for what he could not help, it must remain shut against all mankind, as he was not able to see the mighty difference in men that their hopeless separation implied; — a separation inconceivable to the vast majority of sincere believers in a future state, — believers in Christ, and heirs to heaven under His testament.

Only saw the reflection.

A separation inconceivable.

Their religion — the religion of the people — was not a science nor a profession; it was a life; dogmatic theology was not a part of it. It did not consist in words, but in spirit. Its essence was in the Sermon on the Mount, and in the New Commandment. Love was its ruling principle. God and humanity was their unwritten creed. It taught reverence of the Creator, and charity for the creature. Humility and amity were its fruits. They loved God, and trusted Him; there was not, to them, a single element of terror in His attributes; Indulgent Parent was the language they most used in addressing Him. When they prayed, they used not vain repetitions; their Father knowing what things they had need of before they asked Him. Rarely other prayer than the Lord's was made use of — the sum and summary of all adoration and supplication. They did not disfigure their faces by assuming sad countenances; they did not toss up their eyes sanctimoniously. Confidence in the promises made them tranquil and grateful; they reposed in them. Sound morality was a great part of their religion. Moral honesty — integrity to the core — was its chief corner-stone. At

<small>THEIR RELIGION.</small>

<small>*Humility and amity its fruits.*</small>

<small>*They reposed in the promises.*</small>

Substance and Shadow 263

the foundation of the character of every genuine Sub-Cœlumite there were virtues and elements, cemented and established, to make it worthily everlasting. He felt himself continually searched by the eye of Omniscience, and the observation and estimate of the world were of secondary importance to him. He distinguished between the real substance, character, and its shadow, reputation. He was careful about repeating the Lord's Prayer, as he could not help regarding it as a test of himself, as well as an act of adoration to Deity. Before pronouncing the words, Forgive us our debts, as we also have forgiven our debtors, he hesitated, and inquisition began. Conscience donned the ermine, and consciousness testified. Conceit of sanctity was not a natural result of such self-examination. The ideal seemed further from attainment with every effort; but effort was encouraged to become habitual by increased sense of responsibility. An individual, not responsible to party or sect, he had a conscience toward God. Doing his best to live virtuously and walk humbly, he confidently trusted the Creator to take care of the creature. With the highest standards of conduct practicable or at-

Felt himself continually searched.

Conscience donned the ermine.

Confidently trusted.

tainable, he judged himself not less severely than his neighbor. The Golden Rule he believed to be particularly for self-application. His moral anchorages were fixed and habitual. There were things that under no possible circumstances would he do. His principles were in such constant use that they had the look of instincts. His morals were so constantly applied that they had the appearance of habits. As was said, he picked out the marrow of religion, leaving the bones of theology to the professors. Sectarianism existed, but was not emphatic. Differences of opinion could not be serious where there was only one sentiment. If priests and preachers quarreled to the detriment of religion, they were required to get together and understand one another; nor was authority often necessary to separate them; a few hours' contact reconciled them. Extremes and nice distinctions in faith were more and more forgotten or subordinated; and while a common basis was being discovered, it was felt to be wise by the sects to press differences tenderly. Religion was too essential, they said, to cling to any dogma. It looked to better and immutable conditions. Every

Moral anchorages.

Extremes and nice distinctions.

man believed in immortality; and felt, as had been truly said, that he had a right to this belief; that it corresponded with the wants of his nature. To him, the eternal existence of his soul was proved from his idea of activity; that, if he worked on incessantly till his death, nature was bound to give him another form of existence, when the present one could no longer sustain his spirit. Everything, he exclaimed, with a great soul, is prospective, and man is to live hereafter. That the world was for his education was the only solution of the enigma. He inferred his destiny from the preparation. Whatever it is which the Great Providence made ready for him, it must be something large and generous, and in the great style of his works. The future must be up to the measure of man's faculties,—of memory, of hope, of imagination, of reason. In a word, the life, the character, the faith, the aspirations of the Sub-Cœlumite, all united to make him an intelligent, responsible, religious optimist.

Proof of his eternal existence.

Many of the religious denominations had dwindled away, but those that remained showed a considerable degree of vitality. Descended from parents to chil-

Sects and Creeds.

dren, memory and association clung to them tenaciously. Chapels were everywhere in which sectarian doctrines were still taught. Teachers were zealous, and congregations were faithful, but hardly a particle of bigotry survived. Intelligence and charity had made the sects friendly one with another. No attempt was made by either to turn the key of heaven against the rest. Exclusion or monopoly was no longer dreamed of. A hint of it, even, was an offense to Christianity. Creeds were antiquated; new ones were impossible. People generally thought, and thought differently, and could not again be got to agree upon any set of abstract ideas. Godliness was a mystery they did not attempt to comprehend. It was in the endeavor to know the unknowable that creeds had been produced and sects organized. If its teachers, they said, had continually taught the practice of Christianity, and not expended themselves in developing systems of theology, all Christendom would long since have been a united army against Satan. Alas! they exclaimed, when the gloomy and awful theologies become curiosities, how prodigiously ingenious will the intellects of their inventors appear!

Hardly a particle of bigotry survived.

The gloomy and awful theologies.

Also, in addition to the churches or chapels of the different sects, in all the considerable towns there were commodious cathedrals, in which were sittings for all the inhabitants. These cathedrals were especially sacred to Religious Worship, which, to the enlightened and Christian population of Sub-Cœlum, consisted chiefly in Thanksgiving. Anthems were sung, and choruses, of the most exalted and exalting character. Great organs shook the lofty edifices with their joyful and divine harmonies. When thousands of trained voices, led by the great organ, sang,

<small>WORSHIP.</small>

<small>*Thanksgiving.*</small>

Be Thou, O God, exalted high!

it did seem the Deity was lifted up. In these great cathedrals, at a fixed hour, on Sunday, the sects and the people assembled, and together, in one voice, and with one heart, worshiped God.

<small>*The Deity lifted up.*</small>